Imperfect Competition
and Political Economy

Imperfect Competition and Political Economy

The New Trade Theory in Agricultural Trade Research

EDITED BY

Colin A. Carter, Alex F. McCalla, and Jerry A. Sharples

Routledge
Taylor & Francis Group

LONDON AND NEW YORK

First published 1990 by Westview Press, Inc.

Published 2018 by Routledge
52 Vanderbilt Avenue, New York, NY 10017
2 Park Square, Milton Park, Abingdon, Oxon OX14 4RN

Routledge is an imprint of the Taylor & Francis Group, an informa business

Library of Congress Cataloging-in-Publication Data
Imperfect competition and political economy : the new trade theory in
   agricultural trade research/edited by Colin A. Carter, Alex F. McCalla, and
   Jerry A. Sharples.
      p. cm.
   Includes bibliographical references and index.
   ISBN 0-8133-7993-8 (sc)
      1. Produce trade—Government policy.   2. International trade.
   3. Protectionism.   4. Free trade.   I. Carter, Colin Andre.
   II. McCalla, Alex F., 1937–   .   III. Sharples, Jerry A.
   HD9000.6.I47   1990
   382'.41—dc20   90-12651
                                                                            CIP

ISBN 13: 978-0-367-01560-2 (hbk)

ISBN 13: 978-0-367-16547-5 (pbk)

Contents

v

PART TWO: The Political Economy of Trade

Preface

The papers in this book were presented at a symposium in Montreal sponsored by the International Agricultural Trade Research Consortium (IATRC) in July 1989.

The IATRC is a group of economists from around the world who are interested in fostering research and providing a forum for the exchange of ideas relating to international trade of agricultural products. Each summer the IATRC sponsors a symposium on a topic relating to trade and trade policy from which proceedings are published. For a list of past symposia and related publications, contact Laura Bipes, IATRC Administrative Assistant, Department of Agricultural and Applied Economics, University of Minnesota, St. Paul, MN 55108, United States of America.

Thanks go to Laura Bipes and Don McClatchy for making local arrangements and managing the many details of the symposium. Special thanks goes to Nancy Ottum for editorial assistance and preparation of the camera ready copy of this book.

Colin Carter
Alex McCalla
Jerry Sharples

1

Introduction

Colin A. Carter and Alex F. McCalla

Few bodies of economic theory have held sway longer than the notions of comparative advantage and the neoclassical proof of the gains from free trade. This theory has many simplifying assumptions, such as, homogeneous products, constant returns to scale, perfect competition, and "small" countries. The theory predicts that countries will export those goods that use the country's abundant factor most intensively and that it will import those goods which do not. It suggests a pattern of trade primarily characterized by inter-industry trade (i.e., countries might export wheat and import automobiles). But what we observe in the real world is a trade characterized more and more by intra-industry trade (i.e., countries export and import similar products). For example, the United States both exports and imports beef in large volumes.

There have been significant new and innovative developments in trade theory in the past few years. This work has been recently surveyed and discussed by Krugman.[1] Recent developments have attempted to do a better job of explaining intra-industry trade flows.[2] There are two major streams of new thinking.[3] The first incorporates imperfect competition into trade theory and questions the conventional wisdom derived from neoclassical trade theory, that free trade is always best. It relaxes traditional assumptions of constant returns to scale, homogeneous products and competitive markets. This allows the theory to include increasing returns, differentiated products and strategic trade policy. The second stream of thought assimilates the theory of public choice into trade theory. Unlike the case with imperfect competition, the "political economy" approach to modelling trade is consistent with a free trade approach. Figure 1.1 lists some of the names of major contributors to these two streams of theory.

These new ideas in trade theory have potentially important and widespread implications for agricultural trade research. After all, agricultural trade is characterized by differentiated products, imperfect competition, and strategic trade policy such as export subsidies, import quotas, etc. Virtually none of the agricultural trade modelling to date has incorporated these new theoretical developments. Advances in the theory of the political economy of trade may offer insights into the factors influencing trade barriers. Protectionism has grown in the last decade and the key to reform lies in an understanding of who gains and who loses from protectionism.

These new theoretical developments are not without controversy. For example, Bhagwati has argued (see the reference in footnote 3 above) that there is nothing wrong with the traditional theory, but in the real world trade flows have become distorted by the political behavior of interest groups interacting with non-altruistic governments. In addition, the *Economist*[4] newsmagazine has referred to the new strategic trade theory as the "Next Laffer Curve" because in the magazine's editorial view that strategic trade theory could be used as an excuse for protectionism by policy makers and yield bad policy.

As shown in Figure 1.1, there has already been some work completed on the application of political economy models to agriculture (e.g., Rausser and Freebairn; Riethmuller and Roe; Sarris and Freebairn; Paarlberg and Abbott; Lopez; Miller; Gardner and Baliscan; and Roumasset). These applications have built on the public choice (political economy) work of Baldwin, Caves, and Peltzman and Becker. Bhagwati has categorized the political economy models as either self willed government (S. willed Gov't) or clearing house government (C. house Gov't) models. The self willed government models assume the government possesses a societal welfare function and it chooses policies as though it were maximizing this function. Alternatively, clearing house government models assume the government has no preconceived welfare function but that it only responds to interest group pressures and sets policies accordingly.

Unlike the case for political economy models, there has been little work to date on the application of imperfect competition/strategic trade models. This is surprising given the many potential topics (e.g., targeted export subsidies, managed quotas, etc.). One exception is the work of Marie Thursby (1988) on the application of Brander-Spencer type strategic trade theory to world wheat markets.

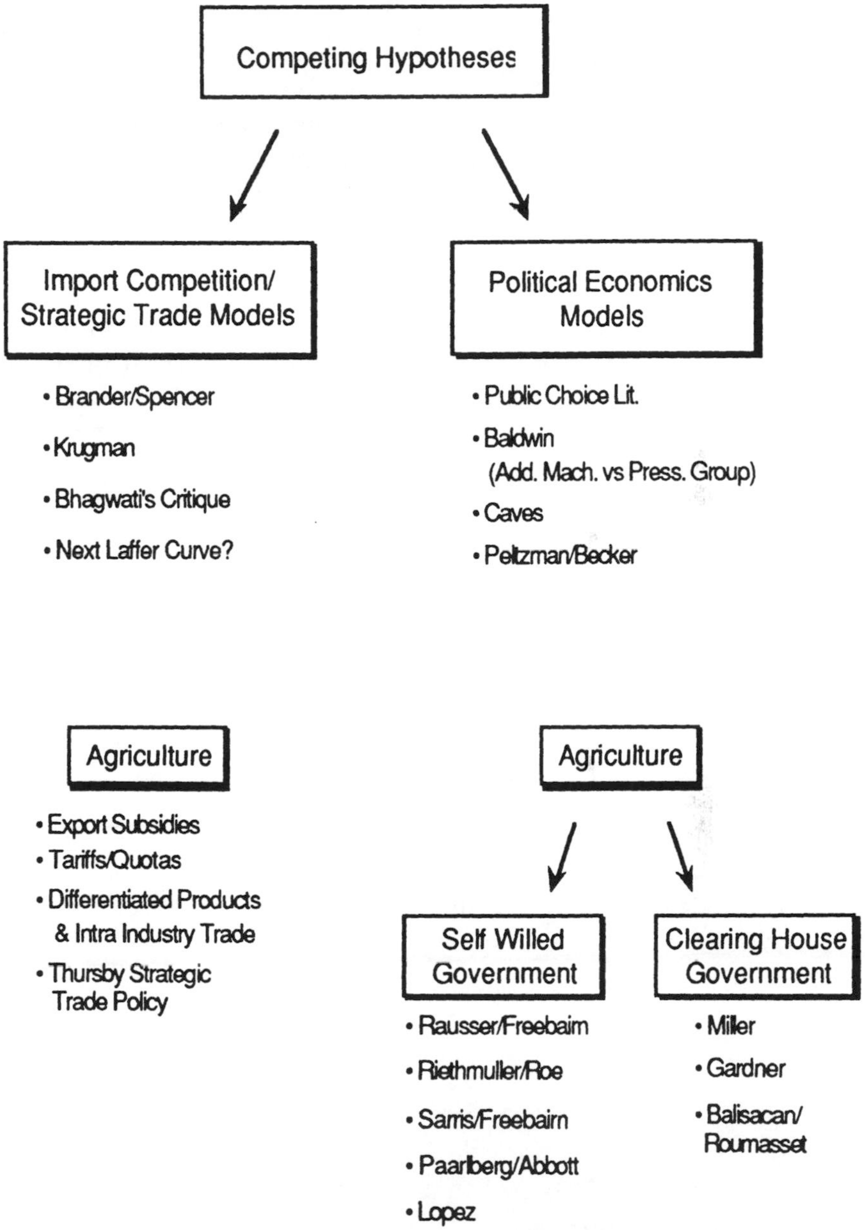

Competing Hypotheses

Import Competition/
Strategic Trade Models

- Brander/Spencer
- Krugman
- Bhagwati's Critique
- Next Laffer Curve?

Political Economics
Models

- Public Choice Lit.
- Baldwin
  (Add. Mach. vs Press. Group)
- Caves
- Peltzman/Becker

Agriculture

- Export Subsidies
- Tariffs/Quotas
- Differentiated Products
  & Intra Industry Trade
- Thursby Strategic
  Trade Policy

Agriculture

Self Willed
Government

- Rausser/Freebairn
- Riethmuller/Roe
- Sarris/Freebairn
- Paarlberg/Abbott
- Lopez

Clearing House
Government

- Miller
- Gardner
- Balisacan/
  Roumasset

Figure 1-1. New Trade Theory

The International Agricultural Trade Research Consortium (IATRC) decided to more fully explore the implications of these new threads of theory for agriculture by sponsoring a symposium on the subject. The IATRC commissioned seven papers which systematically explore the conceptual and empirical dimensions of the new theory and try to determine the potential application to agricultural trade and trade policy analysis.

These papers and the comments of invited discussants are presented in this volume in two parts. Part I addresses imperfect competition issues. Chapter 2, authored by Kala Krishna and Marie Thursby, explores conceptual developments in imperfect competition and strategic trade theory. Chapter 3 by David Richardson is a comprehensive review of empirical research under imperfect competition. Chapters 4 and 5 investigate applications of imperfect competition developments to agricultural markets. Marie and Jerry Thursby (chapter 4) develop a three country model of the wheat market. Donald MacLaren (chapter 5) reviews the implications for modelling imperfect substitutes.

Part II addresses political economy topics. Chapters 6 and 7, by Michael Moore and Edward John Ray, respectively, similarly explore conceptual and empirical analyses on the political economy of trade. Chapter 8 by Harry de Gorter and Yacov Tsur develops a model of political economy of agricultural trade policy. The final chapter by Stephen Haley and Jerry Sharples provides a synthesis and summary of the conference.

These chapters, which include some very insightful comments by invited discussants, represent a major collection of knowledge in a single location, thus allowing agricultural trade researchers to delve into these new developments and apply those relevant parts to their own analysis.

Notes

1. Paul R. Krugman. 1987. "Is Free Trade Passé?" *Economic Perspectives.* 1:131-44.

2. For example, see the collection of papers in Robert C. Feenstra, ed. 1988. *Empirical Methods for International Trade.* Cambridge: MIT Press.

3. Jagdish Bhagwati. "Is Free Trade Passé After All?" Acceptance speech on the occasion of the award of the Bernhard Harms Prize at Kiel, Germany (June 25, 1988).

4. See "The Dangerous Science" in the *Economist,* June 24, 1989, p. 19.

References

Baldwin, Robert E. 1985. *The Political Economy of U.S. Import Protection.* Cambridge: MIT Press.

Baliscan, A. and J. Roumasset. 1987. Policy Choice of Agricultural Policy: The Growth of Agricultural Protection. *Weltwirtschaftliches Archiv.* 123:232-248.

Becker, G. 1983. A Theory of Competition Among Pressure Groups for Political Influence. *Quarterly Journal of Economics.* 98:371-401.

Bhagwati, J. Is Free Trade Passé After All? Acceptance speech on the occasion of the award of the Bernhard Harms Prize at Kiel, Germany, June 25, 1988.

Caves, R.E. 1976. Economic Models of Political Choice: Canada's Tariff Structure. *Canadian Journal of Economics.* 9:278-300.

Miller, T.C. 1989. Explaining Agricultural Price Policies Using a Model of Competition Between Interest Groups. Mimeo, Baylor University.

Paarlberg, P.L. and P.C. Abbott. 1986. Oligopolistic Behavior by Public Agencies in International Trade: The World Wheat Market. *American Journal of Agricultural Economics.* 68:528-542.

Peltzman, Sam. 1976. Towards a More General Theory of Regulation. *Journal of Law and Economics.* 19:211-240.

Rausser, G.C. and J.W. Freebairn. 1974. Estimation of Policy Preference Functions: An Application to U.S. Beef Import Quotas. *Review of Economics and Statistics.* 56:437-449.

Riethmuller, P. and T. Roe. 1986. Government Behavior in Commodity Markets: The Case of Japanese Rice and Wheat Policy. *Journal of Policy Modeling.* 8:327-349.

Sarris, A.H. and J. Freebairn. 1983. Endogenous Price Policies and International Wheat Prices. *American Journal of Agricultural Economics.* 65:214-224.

Thursby, Marie. 1988. Strategic Models, Market Structure and State Trading: An Application to Agricultural Markets. In R.E. Baldwin (ed). *Trade Policy and Empirical Analysis.* Chicago: University of Chicago Press.

PART ONE

Imperfect Competition and Trade

2

Trade Policy with Imperfect Competition: A Selective Survey

Kala Krishna and Marie C. Thursby

Introduction

A vast literature exists on international trade policy with imperfect competition. This literature has generated a lot of interest in policy circles as it is seen as providing a rigorous justification for interventionist policies. There is a plethora of models with an often confusing and apparently conflicting set of associated results. This paper is a selective survey in which we ask what the major contributions of this literature are for policy purposes. In particular, attention is paid to how these contributions might be applicable to international trade issues in agriculture. Given the attention paid to agriculture in the ongoing GATT negotiations, such a survey is timely; and its potential relevance is clear given the existence of marketing boards and increasing returns in transportation and/or processing agricultural commodities.

It is not our intention to do an exhaustive survey of this literature here. Rather, we plan to focus on whether there are any unified themes emerging from this growing body of work. To do this we characterize policies by whether they are "indirect" policies, such as taxes/subsidies, or "direct" policy instruments, such as export restraints (ERs)[1] or ceilings or floors on prices which affect the strategic environment. As noted above, we shall restrict our attention to models which we feel are most applicable to the institutions which exist in agricultural marketing. We shall also restrict our attention to partial equilibrium

9

models as this is where the bulk of the work, both theoretical and empirical, has been done.

As there are a number of excellent surveys of trade policy with imperfect competition (for example, Grossman and Richardson (1985), Dixit (1984), (1987) and Venables (1985) as well as others not mentioned here), a natural question to ask is what warrants doing another survey. Such a survey is desirable for two reasons. First, the effects of both indirect and direct policies depend greatly on market structure. A number of issues either do not arise or take different forms depending on the market structure. For example, issues concerning the impact of export restraints on collusion among producers do not arise by definition in competitive markets. Questions related to the supply response that occurs due to the entry and exit of firms takes a different form depending on the market structure. This is not the focus of existing surveys. Second, there has been a lot of work in this area in recent years. There is a profusion of models and results and it is hard to put the often seemingly contradictory results in perspective. Although other surveys have contributed in this regard, they have not focused on the special characteristics created by agricultural institutions.

The work on trade policy in imperfectly competitive markets is explicitly game theoretic in character. This is because imperfectly competitive markets firms are "large" and so take account of the fact that their actions affect the market and the actions of other agents. Thus all the interesting aspects of strategic behavior arise in these models. These include, but are not limited to, the use of threats and their limitation by their credibility, and the choice of a variety of instruments to precommit for strategic reasons.[2] These aspects have no place in perfectly competitive environments. Such behavior may have unexpected effects or expected effects for unexpected reasons. Intuition based on standard competitive models can therefore be seriously flawed.

In the next section we present a stylized overview of the areas we will survey. The following section discusses the use of indirect policies and the lessons to be garnered for agriculture from this literature. As will be argued in the stylized overview, direct restrictions have very different effects from indirect ones. Section 4 discusses the literature in this area. Section 5 concludes.

A Stylized Overview

A useful way to organize an examination of the effects of indirect and direct policies is outlined in equations (1) - (5). First, consider the competition of n firms (some foreign and some domestic) in the absence of trade policy. While firms may well have several strategic variables, we shall focus on the case where there is a single strategic variable.[3] Equation (1) states that the profit function, π^i, of the ith firm in the absence of trade policy depends on the values taken by its own strategic variable, S^i, as well as those of its competitors, denoted by S^{-i}:

$$\pi^i(S^1,...,S^n) \quad i=1,...,n. \tag{1}$$

Each firm chooses the level of its strategic variable to maximize its profits, taking S^{-i} as given. The levels of the strategic variable which are mutually consistent in the absence of trade policy give the Nash equilibrium under free trade and are denoted by $S^1(F),..., S^n(F)$. Given that the other firms are choosing these levels of their strategic variable, in the Nash equilibrium each of the n firms is best off choosing $S^i(F)$. This gives equilibrium profits with free trade:

$$\pi^i(S^1(F),...,S^n(F)) \quad i = 1,...,n. \tag{2}$$

for each of the n firms. The equilibrium levels of other endogenous variables, such as welfare can be derived from the equilibrium levels of the strategic variables and the specification of the model.

Now consider the effect of introducing tax/subsidy policies. Each firm's profits, with an indirect policy set at t, are given by:

$$\pi^i(S^i,...,S^n,t) \quad i = 1,...,n. \tag{3}$$

The equilibrium values of the strategic variable are given by $S^1(t),...,S^n(t)$, and equilibrium profits are given by $\pi^i(S^1(t),...,S^n(t),t)$ for i = 1,...,n when the indirect policy is set at t. If t is a tariff,[4] the free trade level of the indirect policy is $t = t^f = 0$. However, $\pi^i(S^1,...,S^n,t^f) \equiv \pi^i(S^1,...,S^n)$. That is, when $t = t^f$, profits are unchanged for *all* values of $(S^1,...,S^n)$, so that the equilibrium is the free trade equilibrium, and equilibrium profits are given by (2).

One strand of the literature has focused on how the equilibrium is altered by governments setting these policies strategically to maximize their objective functions. For several reasons, the policy implications of this type of analysis differ from those with perfect competition. One reason is the ability of firms to act strategically. In addition, imperfect competition gives rise to the possibility of positive economic profits in equilibrium, and these profits enter the objective functions of governments. Finally, by moving first, governments can precommit to policies which firms take as given and which change the equilibrium of the game firms play.

The levels of indirect policies which maximize the government's objectives and the resulting equilibrium outcomes are dependent on special characteristics of timing (i.e., who moves when), the strategic variable used by firms (for example, price or quantity), a well as other characteristics of market structure, such a entry and exit. Not surprisingly, the results are also dependent on the set of instruments available to the government. It is important to consider a full set of instruments as otherwise policies that seem optimal may be inappropriate.

Direct policies have different effects because they alter the *profit function* of a firm even when the levels of the constraints imposed by the policy are set at free trade levels. If the free trade equilibrium strategies are played, profits are free trade profits for all firms. However, if other strategies are played, profits need not be what they would have been given these strategies and free trade. Hence, the profit *function* changes. For example, consider an export restraint. For simplicity, take the case with two firms, one foreign and one domestic, when each firm has only one strategic variable, price, and the export restraint set is at the free trade level. To see that the profit functions will differ from those under free trade in the absence of a restraint, consider what happens if a foreign firm charges the free trade price. In this case, the foreign firm's profits will be different from those under free trade because its demand exceeds the level of the export restraint. The domestic firm's profits will be different because the foreign firm's consumers are being rationed at these prices and some of them will choose to buy the domestic product. The profits at the old free trade equilibrium are unchanged, but the change in profits out of equilibrium can alter the choices of the firms and hence alter the equilibrium.

The same reasoning also applies with more firms, so that with n_1 home firms and n_2 foreign firms, export restraints need not be restrictive for them to have an effect, as their presence changes the relevant profit function for each firm from $\pi^i(.)$ to $\Pi^i(.;V)$. Thus we denote the profits of the ith firm with an export constraint of V in effect by:

$$\Pi^i(S^1,...,S^n;V) \quad i=1...n \tag{4}$$

where $n = n_1 + n_2$. Notice the function itself has changed even when $V = V^f$, its free trade level. With the export constraint, firms choose their strategic variables to maximize the relevant profit function, $\Pi^i(.,V)$. Again, the levels of these strategic variables which are profit maximizing and mutually consistent are the Nash equilibrium ones. The equilibrium profits of the firms are then given by:

$$\Pi^i(S^1(V),...,S^n(V)) \quad i=1...n. \tag{5}$$

and the values of the other endogenous variables are a function of the equilibrium levels of the strategic ones as before.

This discussion highlights the fact that the policy implementation procedure is crucial as is the choice of strategic variables allowed. The former determines the nature of the $\Pi^i(.;V)$ functions while the latter affects the equilibrium values of the strategic variables actually chosen. We implicitly assume that variables not taken to be strategic do not change with the imposition of an export restraint.[5]

Indirect Policies

In this section we focus on the major ideas to emerge from the analysis of tax/subsidy policies with imperfect competition. We begin by considering work done in the simplest model of imperfect competition, monopoly. We then move on to more complicated models of static oligopoly. Here the timing structure in the oligopoly game is important as the results are sensitive to the assumptions made. As we shall see, the government's ability to act as a first mover gives rise to many of the policy implications here. Another crucial aspect is the choice of strategic variable. This does not arise in models of monopoly; whether price or quantity is chosen is irrelevant in models of monopoly but it is vital in models of oligopoly.[6]

Monopoly

The starting point for much of the analysis of trade policy with monopoly is a paper by Brander and Spencer (1981) on the extraction of foreign monopoly rent. Their point is a simple one. When imports are supplied by a foreign monopolist, the price paid by domestic consumers exceeds the monopolist's marginal cost. This gives rise to the potential use of trade policy to extract foreign monopoly rent. This point was made earlier by Katrak (1977), but Brander and Spencer make the important point that potential entry of domestic firms affects the appropriate policy.

For the simplest case, suppose imports are produced by a foreign monopolist and some entry barrier prevents domestic entry. Figure 2-1 replicates Katrak's analysis of a monopolist facing linear demand and constant marginal cost. With free trade Q_m is imported at price P_m. To see that a tariff can raise domestic welfare, consider the effect of a specific tariff of t per unit. Imports decline to Q_t and their price rises to P_t. Domestic consumer surplus is lower by the hatched area, while tariff revenue is given by the dotted area. For a small tariff the cross-hatched triangle can be ignored, so that welfare rises if the tariff revenue exceeds the single-hatched area as is the case with linear demand and constant marginal cost.

Even without potential entry, this result is sensitive to the demand and cost structure. Suppose the demand is a constant elasticity function. A specific tariff decreases consumer surplus by more than the tariff revenue as the marginal revenue curve is flatter than the demand. In this case a specific subsidy improves welfare both at home and abroad. With constant cost, the price decreases by more than the per unit subsidy so that consumer surplus increases by more than the cost of the subsidy. However, this is hardly rent extraction since the foreign monopolist's profits increase. Brander and Spencer (1984b) make this point, as well as deriving appropriate policy when tax instruments are ad valorem.

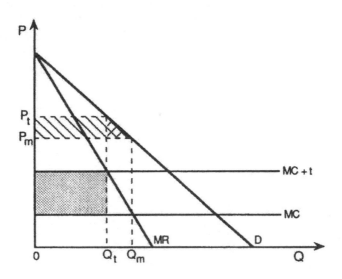

Figure 2-1. Rent Extraction with Foreign Monopoly

To see the effect of potential entry, consider a Stackelberg leader model where the foreign firm is the leader and a potential domestic entrant takes as given the output of the leader.[7] This is the situation analyzed by Brander and Spencer (1981). There is a fixed entry cost for the entrant, so that above some level of imports, profits of the entrant are nonpositive. Figure 2-1 is drawn assuming this level is lower than the monopoly level of imports, Q_m. Now suppose the domestic entrant can earn positive profits if the foreign firm produces Q_m, but that the entrant's profits become zero at some higher level of the foreign firm's output, Q_d. In this case, the foreign firm will compare the profits from producing $Q_d > Q_m$ with the profits it would make in the equilibrium of a Stackelberg leader game. We shall refer to the former as entry deterrence profits and the latter as Stackelberg profits. If deterrence profits are higher, it will produce Q_d; otherwise, it will act as a Stackelberg leader.

Several interesting results emerge. First, when entry deterrence is the outcome, a tariff will extract some foreign rent without a cost to domestic consumers. This is because the tariff does not alter the minimum output required to deter entry. The foreign monopolist will continue to sell Q_d at $P(Q_d)$ as long as its deterrence profits exceed Stackelberg profits. Second, the tariff lowers profits associated with both the entry deterring and Stackelberg outcomes. An increase in the tariff lowers the Stackelberg profits less than those with entry deterrence. This means a sufficiently high tariff will induce entry by the domestic firm.

Before turning to oligopoly, we note the sharp contrast between these results and those under perfect competition. Under perfect competition a tariff would improve domestic welfare only if it changed the foreign price (= marginal cost), and a subsidy would not improve welfare net of its cost.

Oligopoly

We begin by considering the simplest models where policy prescriptions are the clearest. We restrict our attention to static models, and only the home government is assumed active in policy. As does most of this literature, we assume the government acts first. It chooses the levels of policy instruments to maximize welfare, considering the effect on firms' decisions. Firms move second (but simultaneously) and make their decisions taking government policy as given.

Brander and Spencer (1985) consider a duopoly model in which a foreign and home firm produce a homogeneous good for a third market.

Neither firm produces for its domestic market, so that issues related to domestic consumer surplus do not arise. Home welfare is the firm's profits net of government revenue. Firm rivalry is modelled as a Cournot game, and they show that optimal policy for the home government is to subsidize exports.

Figures 2-2a and 2-2b are useful in explaining this result. Figure 2-2a depicts the best response functions of the home and foreign firms, $B(Q^*)$ and $B^*(Q)$, respectively. Demand is assumed to be linear, marginal cost is constant, and quantity is the strategic variable. With Cournot competition, each firm chooses its output level (in this case, exports) to maximize its profits, taking as given its rival's output (exports). The Nash equilibrium with free trade is given by point C. The home firm's free trade profits are denoted by π^C. Notice the firm could earn higher profits if it could precommit to a higher level of exports, but such a commitment is not credible in the absence of policy. Since the government moves first, it can set policy to shift the best response function in the appropriate direction. A per unit export subsidy (or producer subsidy, as they are equivalent here) will shift the home firm's best response function out. The optimal subsidy shifts it to $B(Q^*,s)$ which gives the home firm Stackelberg profits in equilibrium.

An alternative explanation is in terms of perceived and actual marginal revenue, as in Helpman and Krugman (1989). Each firm chooses its output so as to equate its perceived marginal revenue with marginal cost. The firm's perceived marginal revenue is based on the assumption that its rival's output is given. With quantity as the strategic variable, an increase in home firm output will decrease price by less than is perceived by the home firm. This is because foreign output decreases along its best response as home output increases. Hence the actual marginal revenue (depicted by MR_a) is flatter than that perceived by the firm, and will intersect marginal cost at a higher output than the free trade level. The optimal subsidy induces the firm to produce this higher output level, Q_s. In essence there is a strategic distortion in this model which a first mover government can offset by appropriate policy.

Eaton and Grossman (1986) make this point and show the strategic variable is critical in determining the distortion. With Bertrand competition, for example, they show the appropriate policy is an export tax. Figure 2-3a replicates the best response functions of the home and foreign firm in Eaton and Grossman's analysis. These best response functions have the usual properties when price is the strategic variable

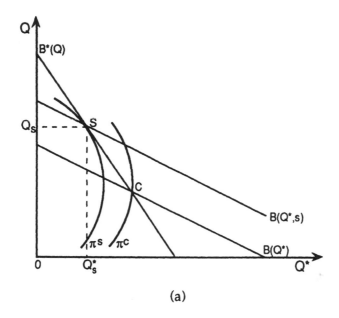

(a)

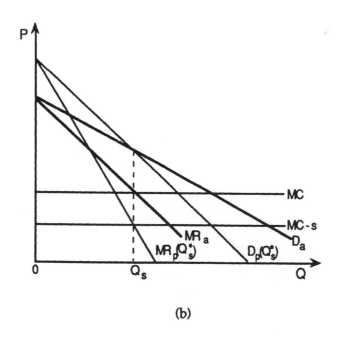

(b)

Figure 2-2. An Export Subsidy with Cournot Competition

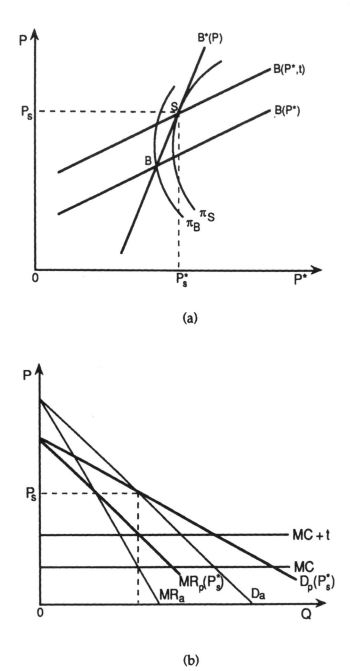

(a)

(b)

Figure 2-3. An Export Tax with Bertrand Competition

and the goods are substitutes. With Bertrand competition, each firm chooses price to maximize its profits, taking as given its rival's price. With free trade the equilibrium is given by B, and the home firm's profits are noted by π_B. Home firm's profits are highest given B*(P) if it can commit to p_s. An export tax per unit of t achieves this. Figure 2-3b illustrates why the appropriate policy is a tax. Perceived marginal revenue is based on a given foreign price, while actual marginal revenue incorporates foreign response as seen by a first mover. A government which can precommit to policy would optimally set policy to equate this actual marginal revenue with marginal cost. The optimal policy with Bertrand is a tax, because here the perceived marginal revenue is flatter than the actual. That is, if the home firm thinks of raising price, it perceives a larger decrease in demand than occurs. This is because it does not account for the rise in price along the foreign best response as home price rises. And export tax corrects this strategic distortion, inducing the home firm to choose P_s.

When there is more than one home firm, there is an additional distortion created by the fact that each firm does not internalize the effect of its decisions on the terms of trade. This is pointed out by Dixit (1984) and Eaton and Grossman (1986). The terms of trade distortion calls for a reduction in exports from the free trade level. This distortion works against or reinforces the strategic motive for policy, depending on the nature of competition. With Bertrand competition, it reinforces the case for an export tax, while it works against the strategic motive for an export subsidy with Cournot competition. A treatment of the latter is in Krishna and Thursby (1988). Horstmann and Markusen (1986) examine policy with increasing returns and free entry. In their model of Cournot competition, interventionist arguments collapse to terms of trade arguments.

Policy arguments are also more complicated when producers sell in their own markets as consumer surplus becomes an issue. Depending on the number of firms, there may be strategic, terms of trade, and consumer distortions all affecting optimal policy. This has been pointed out in several contexts (Brander and Spencer (1984a), Dixit (1984, 1986), Eaton and Grossman (1986), Krishna and Thursby (1988), and Thursby (1988), to name a few).

A targeting approach to policy is developed in Krishna and Thursby which helps explain the nature of optimal policies when there are multiple distortions.[8] This work is of particular interest here as it is one of the few papers in this literature to focus on agricultural institutions.[9] In this paper production is perfectly competitive, but producers (farmers) do not sell directly to consumers. They sell their output to large marketing agents or boards who then sell to consumers at home and abroad. It is shown how policies are affected by the objectives of these boards and whether or not they are regulated, as well as arbitrage possibilities and the number of boards competing. The government is assumed to be able to tax or subsidize exports, domestic production, and consumption.

To illustrate the targeting principle, consider the simplest case of duopoly with market segmentation. A home and foreign board compete in a third market, but each is the sole supplier to consumers in its own market. There are three distortions possible. The boards have monopoly power over domestic consumers. They may or may not exercise monopsony power over competitive suppliers depending on whether the boards are producer cartels or monopsonists. And finally there is a strategic distortion, which calls for different policies depending on the strategic variable. Because of this, we adopt a conjectural variations[10] approach to allow for either Cournot or Bertrand competition. The monopoly and monopsony distortions are optimally targeted by production and consumption subsidies, and trade policy optimally targets the strategic distortion. Whether the appropriate trade policy is a tax or subsidy, as usual, depends on the strategic variable.

Our discussion so far is based on the assumption that boards are not directly regulated. Marketing boards are often regulated in their domestic pricing, and this will be reflected in the targeting principle. A regulation which enforces marginal cost pricing at home, for example, links distortions by linking domestic and export sales. While the regulation removes the consumption distortion, it creates another distortion since it encourages boards to raise exports in order to raise marginal cost and domestic price. Optimal policy can be implemented by a single instrument, a trade tax/subsidy, and its level is determined by both the linkage of distortions and the strategic distortion.

Krishna and Thursby also consider the effect of arbitrage on targeting.[11] With arbitrage, a board cannot determine the amount of home and foreign sales independently. In the absence of regulation, this means that the consumption (monopoly) distortion is linked through arbitrage to the price abroad and the strategic distortion includes a linkage effect as well. With arbitrage, a regulation of domestic price eliminates any strategic distortion and a role for policy exists to the extent that the terms of trade can be affected by policy. Essentially arbitrage and price regulation together force the board to act much like a competitive board. This suggests that a useful way to think of policy comparisons in situations with and without arbitrage is in terms of a regime change.

All of these studies assume the government sets the level of policy instruments before firms make their decisions. Gruenspecht (1988) alters the timing structure, and shows that an export subsidy can increase profits of Bertrand competitors. In his model the government establishes a subsidy program, but firms choose their prices before the subsidy levels are set. The government sets these to maximize profits net of weighted subsidy cost. The profit enhancement occurs because the subsidy regime itself acts like a facilitating device.

Recent work has also examined models with trade in intermediate inputs as well as final goods. Spencer and Jones (1989) and Rodrik and Yoon (1989) analyze policy with integrated firms. In both studies an integrated firm chooses whether to export an intermediate input to its foreign rival (in the final good market) or to vertically foreclose (i.e., to refuse to supply the input to its rival). And in both studies, the integrated firm has access to cheaper technology for the input and acts like a Stackelberg leader in the input market. In the final goods market the firms act as Cournot rivals. Spencer and Jones focus on policy from the perspective of the integrated firm's government. Their results are quite different from the results for simpler models with Cournot competition. For example, an export tax (rather than subsidy) is optimal in the final goods market (as well as the input market). Rodrik and Yoon consider policy from the importing country perspective. A tariff on the intermediate input is borne entirely by the foreign firm,[12] as the vertically integrated firm charges the input price which just prevents the domestic firm from developing the input itself. If the import country government subsidizes the domestic firm's development of the input, welfare is increased because the foreign rival will lower the input price and the subsidy will never be dispersed in equilibrium. A tariff on the final good will also increase welfare.

Direct Policies

Under this heading we will discuss the main ideas that emerge in the analysis of direct policies using export restraints as the illustrative example. As in the previous section, we begin by considering work done in the simplest model of imperfect competition, monopoly. Then we move on to slightly more complicated models of static oligopoly. As is true with indirect policy, the timing structure in the oligopoly game is important as results are sensitive to the assumptions made. To highlight this we discuss the work that uses Stackelberg leadership models, where one firm moves first, as well as models where both firms move simultaneously. As before we also emphasize the importance of the strategic variable. Finally we move on to dynamic models where the game is repeated many times or where there is an explicitly dynamic aspect of behavior that is affected by an export restraint.

Monopoly

The starting point for much of the work on export restraints in imperfectly competitive markets comes from the much cited paper, Bhagwati (1965), on tariff quota non-equivalence.[13] The basic point made by Bhagwati is that with domestic monopoly and foreign competitive supply a quota at the free trade level causes prices to rise and domestic consumption to fall. Thus, a quota need not be set at restrictive levels for it to have profound effects. This is in contrast to the competitive case where a quota must be set below the free trade level for it to have any effect. Call this the "M" for monopoly effect.

The logic behind the result is that the residual demand curve facing the monopolist becomes steeper for price increases in the presence of a quota. This occurs because foreign supply cannot expand as price rises when a quota is imposed. This kink in the demand curve makes it profitable for the monopolist to raise his price. If the ER is set below the free trade level there is a further effect because of the ER being set at a restrictive level. Making the quota more restrictive increases the sales of the monopolist with demand that is approximately linear, but reduces total supply, thereby raising price in the domestic market. This effect of a more restrictive ER is common to both monopoly and to models of competition. Call this the "C" effect as it exists even under competition.

One might ask what the model of domestic monopoly implies for the effect of an ER on domestic production. Oddly enough, under domestic monopoly it need not have any effect on it for a wide range of the ER. To see why consider the model of domestic monopoly where the world price is given and the monopolists marginal costs are increasing. This model also illustrates the "M" effect of a quota in the special case where world supply is infinitely elastic. These conditions are depicted in Figure 2-4. CD is the domestic demand curve, DMC is the domestic marginal cost curve. Under free trade the residual demand curve facing the monopolist is given by ABD. The monopolist then produces AE and EB is imported.

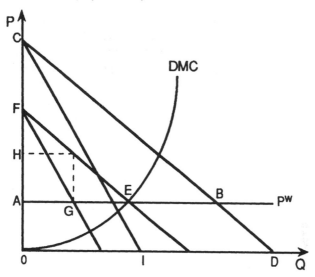

Figure 2-4. An Export Restraint with Domestic Monopoly

Now consider an ER at EB level of imports. The residual demand now is given by FEBD at home and AP^W abroad. The ER lets the price at home exceed the price abroad so that the monopolist can price discriminate in this direction. Equating the horizontal sum of marginal revenues with marginal costs, FGP^W with DMC, gives that AE remains the production level, but only AG is sold at home at price OH, which exceeds OA, while GE is sold abroad at price OA. The ER creates market power where none existed. With an upward sloping foreign supply curve it just enhances it. Varying the ER from being prohibitive to being irrelevant does not affect domestic production in Figure 2.4 as DMC intersects P^W to the right of CI, the marginal revenue with a zero quota. Had this intersection been to the left, domestic output would start rising as the quota became very restrictive but would be unchanging till then.

In essence this result emerges because the effective marginal revenue curve remains flat at the world price despite changes in the quota. This fixes domestic output at the point at which DMC equals the world price. Since this point does not change, only the allocation of the output across markets changes. This should be contrasted to the case of competitive supply where reducing the quota raises domestic prices which in turn elicits a greater domestic supply as the quota level falls.

Oligopoly

With oligopolistic markets it is important to carefully model the effects of a VER on the game played by the firms. Duopoly models are traditionally used here. Stackelberg leadership models show that quotas even imposed at the free trade levels have effects. The idea is very simple. An ER alters the best response functions. Even if these are not affected at the free trade levels but elsewhere, this affects the equilibrium in a Stackelberg leadership model as the leader maximizes along the followers best response function.

As an illustration consider the duopoly case when the home firm is the leader. Under free trade the equilibrium is given by the point S in Figures 2.5a and 2.5b which correspond to price and quantity being the strategic variables. The case with linear demand and constant marginal cost is depicted for illustrative purposes. A quota at the free trade level alters the foreign best response function. In Figure 2-5a the line P'P traces out the set of prices where the free trade level of imports are demanded. The best response function of the foreign firm with an ER at the free trade level is given by the dark line. It is the free trade best response when this calls for a higher price than along P'P and is P'P when the opposite holds. In the former case the foreign firm is not constrained by the quota, while in the latter case it is. The Stackelberg equilibrium with a quota is at S'. Here the firm has higher profits than at S. The foreign one also has higher profits as it sells the same amount at a higher price. Both firms gain from a quota. By the use of continuity arguments[14] the result goes through for ERs close to this level as well. Similar results obtain with quantity competition. Again the dark line gives the new best response function of the foreign firm, and the equilibrium moves from S to S' in Figure 2-5b. The leader's profits rise as he chooses a new point when the old was available, while the follower's rise as total output falls and so price rises and he sells no less.

This is pointed out in Itoh and Ono (1982) and (1984). They also claim that if the foreign firm is the leader this kind of effect does not occur. This is not quite correct as they *assume* that the home firms best response is not affected by a quota on the foreign firm. However, this is not so unless it is assumed that shortages in the market for one good do not affect the demand in the market for the other, which is inconsistent with the goods being substitutes for on another as is being assumed. This is pointed out in Krishna (1989b) and noted in Itoh and Ono (1984). This assumption is one of "no spillovers."

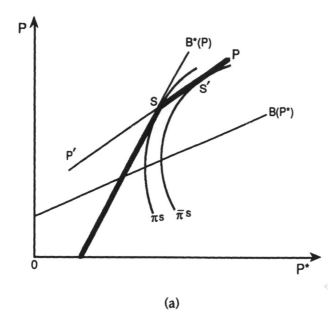

(a)

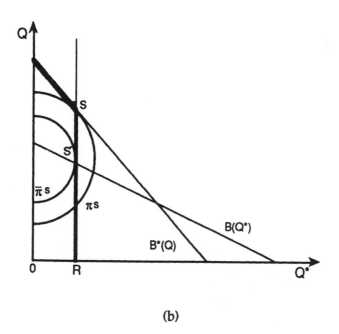

(b)

Figure 2-5. An Export Restraint with Quantity Competition

If a constraint on the foreign firm exists, the home firms best response function changes as it has the option of making the constraint bind on the foreign firm by choosing to charge a high price for its own good. It uses this option strategically. If the foreign price is low it chooses to charge a fixed high price and create excess demand in the market for the foreign good. If the foreign price is high enough it is best off ignoring the existence of the ER and here its best response is unaffected.

Itoh and Ono (1982) also consider the endogenous determination of leadership and argue that this works in favor of the home firm being the leader. However they make the same assumption here which flaws the analysis. Thus their results are unlikely to hold in general. It remains unclear whether any such result holds when the model is carefully specified and this extension needs doing. In any event it is not clear why one should assume that one firm should be the leader. What allows one firm to move first?

Itoh and Ono (1984) and Harris (1985) analyze the effects of an ER in a duopoly model with one foreign and one domestic firm where neither is a leader. Firms are assumed to compete in price and to produce differentiated substitute products. Itoh and Ono argue that an ER at the free trade level has no effect. However, since they make the no spillovers assumption in deriving this result their result is not general. Harris argues that ERs have an effect by their very existence. However he argues that the ER makes the home firm into a Stackelberg leader. This is what drives his result.

In Krishna (1989b) it is argued that the ER gives the domestic firm the profits of a Stackelberg leader in *equilibrium* when neither moves first. The essence of the argument is that there is a third effect of an ER in oligopolistic markets in addition to the "M" and "C" effects mentioned above. Call this the "I" for Interaction effect. It arises because each agent is affected by the ER and the equilibrium is affected by the interaction of these agents. While the "M" effect can only raise the agents profits the "I" effect can raise or lower them. With an ER it tends to raise them. However with other restraints it can lower them as shown in Krishna (1987).

Consider what happens when an ER is imposed at the free trade level. Why is the free trade price not an equilibrium? At these prices the existence of the ER makes the demand curve facing the home firm less elastic for price increases as a higher price makes the ER bind on the foreign firm. This is the analogue of the "M" effect. This price increase by the home firm makes it profitable for the foreign firm to raise its price as it is effectively supply-constrained by the ER at the free trade price. This is the essence of the "I" effect. The equilibrium is a mixed strategy which gives the domestic firm the profits of a Stackelberg leader when a natural rationing rule is used. It also raises foreign profits. This result depends on whether goods are complements or substitutes. Krishna (1989c) shows that when the products are complements, an ER has no effect on equilibrium if imposed at the free trade level.

Both price and quantity competition models and their variants are used in studying the effects of ERs. However, it is a bit unsatisfactory to use Cournot models in analyzing quantity constraints since these restrict the strategic variable itself, leaving no room for firms to use the existence of the ER strategically. When quantity is the strategic variable, ERs have no effect by their very existence as no "M" effect occurs and hence no "I" effect can occur.

The effects of ERs have also been studied in the context of repeated games. Their effects here are much richer and harder to predict because there are an infinite number of equilibria. Usually, studies focus on the most collusive equilibrium, and in most cases, only trigger strategies are considered. With trigger strategies, collusive equilibria are maintained by the threat of going to the non-cooperative one shot equilibrium.[15] Rotemberg and Saloner study such a model and conclude that there are two effects at play. An ER reduces the profitability of deviation from a collusive equilibrium as foreign firms cannot sell more than the ER by deviating. However, it also reduces the ability to punish deviators as this is also restricted by the ER. Thus whether more collusive outcomes become possible with an ER or the most collusive outcome possible becomes less collusive depends on details of market structure.

In one case, studied by Krugman and Helpman (1989), the answer is unambiguous. They consider the case where there are a number of domestic Cournot oligopolists and foreign supply is competitive. They show that an ER at the free trade level actually reduces the most collusive outcome that can be sustained. Assuming that this is the outcome that is chosen among the continuum of outcomes that can be sustained and assuming that this outcome is less collusive than that of a perfect cartel. Then this implies that an ER will raise output and lower price! This result is extremely counter intuitive and occurs because in this case only the penalty to cheating is reduced in the relevant region.

Bull (1986) studies another aspect of ERs. ERs are often specified as a quantity per year. They are also allocated on a first come first served basis. This creates incentives for "Quota induced sales games" where firms rush to import in order to obtain the quota. With costs of storage it is not best to import everything at the beginning of the year and fill the quota. This makes the problem a non trivial one. And given the fact that quotas are often filled in the first few months of the year even when they are not very small, the problem is a relevant one.

Drawing Policy Conclusions

What are the main points that emerge from the above discussion of impact effects of an ER? How to they relate to evidence on the effects of such ERs? The main point of the above literature seems to be that ERs need not be set at restrictive levels to have significant effects. This result was robust to a variety of specifications of market structure and the expected effect of such ERs was, aside from the Krugman-Helpman result which they regard as a curiosity anyway, that ERs tend to raise price, and encourage collusive behavior. The question then arises as to how one can separate the restrictive effect of the ER from the induced effect on behavior? Moreover, since the effects of ERs are sensitive to the strategic variable chosen and to market behavior how can the appropriate model be chosen?

Recent work on computable partial equilibrium models that tries to empirically implement the corresponding theoretical work holds out some hope in this area. This work is in its infancy and the results reported should be taken at best as suggestive for this reason. Dixit (1987) develops a static oligopoly model and calibrates it to the auto industry data for years including the years when a VER was in effect. He uses a conjectural variations model and derives estimates for the conjectures in the calibration process. In doing so he bypasses the question of what is the appropriate strategic variable to use. However, conjectural variations models have a number of problems with them, the main one being that they have no extensive form associated with them so that it is unclear what the timing structure is.[16] His results suggest that the ER did make behavior more collusive, especially in the early years of the ER in that it changed the calibrated levels of the conjectural variations in this direction. Dixit assumed that domestic goods were perfect substitutes for one another as were foreign goods though domestic and foreign goods were imperfect substitutes for one another. This meant that behavioral estimates derived were biased towards collusion as behavior had to be more collusive than Bertrand behavior for profits to be positive. Krishna et al. (1989) show that the level of the calibrated behavioral parameters changes in the expected direction, that is become less collusive, when a richer specification allowing for product differentiation is used. Their *direction* over time does not.

Lambson and Richardson (1986) attempt to calibrate a repeated game model to the auto industry as well. Their main result is that it is not inconsistent with the data. They specify a repeated game model which is more sophisticated than that of Rotemberg and Saloner or Krugman and Helpman as they do not restrict themselves to trigger strategies. Punishments more severe than the noncooperative one shot equilibrium can sometimes be credible threats as shown by Abreu (1985). The greater punishments in turn allows greater collusion.

Another point worth making is that the supply responses in such models are often perverse. With domestic monopoly the example in this paper suggested that although production for the home market falls, home production is to a large extent unaffected by an ER. With oligopoly a concern with the use of VERs is that as they are not usually imposed on all suppliers, the ER would raise the supply from unconstrained suppliers and affect total imports from other sources. A simple duopoly model with two foreign suppliers and no home supply or an upward sloping domestic supply curve, with one of the producers subject to an ER is analytically equivalent to the model of Krishna (1983). Applying this would yield the result that the unconstrained suppliers output may rise or fall with an ER on the other firm. This suggests that there may be less of this supply diversion than suggested by competitive models.

Although there are a variety of models existing on the effects of ERs, well specified oligopoly ones are few. More work on developing such models along with ways of empirically implementing them would be of help to policymakers since it is clear that the results of computable partial equilibrium or calibration models are greatly influenced by model specification. It is essential for this reason to have some guidance in model specification from econometric work.

Conclusion

What are the main points that emerge from our discussion of trade policy under imperfect competition? Are the policy prescriptions practically applicable? What are the implications for agricultural policy and research?

First, with both indirect and direct policies the intuition one uses to guide policymakers with perfect competition does not apply, in general, to imperfectly competitive markets. Unfortunately, rather than a simple policy message emerging from this work, all sorts of policy options can be appropriate depending on the relevant model.

Special features such as timing, strategic variable, market segmentation are crucial in determining policy effects.

Second, with direct policies, the main point seems to be that they need not be set at restrictive levels to have significant effects. This result is robust to a variety of specifications of market structure. But as in the case of indirect policies, the effects of direct policies are sensitive to the strategic variable chosen and to the market behavior. They are also sensitive to the way in which policies are implemented. This is important in oligopolistic markets because firms will respond strategically to any scheme, and this may have unexpected policy consequences. For example, Krishna (1988 and 1989a) argues that auctioning import quota licenses is likely to raise little or no revenue when the license market is competitive but the product market is not.

Third, it is apparent from this survey that knowing the appropriate model is crucial for policy purposes. But how do we select the appropriate model? While we have mentioned initial efforts to bring empirical content to this literature, this work is in its infancy. Much more work needs to be done here. In particular, a better mix between calibration and econometrics is needed.

Fourth, with regard to agricultural issues, very little of this work has focused on agricultural institutions. Given the prevalence of marketing boards and large trading firms in agricultural markets (see, for example, Thursby and Thursby (1989)), there is good reason for future research to examine issues of market structure and its policy implications in agricultural markets.

Notes

1. If the ERs are "voluntary export restraints" (VERs), i.e., offered by the exporting country, then it is assumed that the exporting country implements them. Hence, if they are implemented by the use of licenses, the exporting country chooses the allocation scheme. For example, it chooses whether and how to give the licenses away to companies or to sell them. If an ER is not voluntary, it is equivalent to a quota implemented by the importer. The difference between a VER and a quota is who implements the ER. For this reason, it is usually assumed that with a VER the rents of the ER go to the exporter, while with a quota they go to the importer. It is necessary to be clear about the details of implementation of ERs because this can greatly affect the welfare consequences. Here we shall use ERs, VERs, and import quotas interchangeably.

2. By "precommit" we mean the ability to do something that is then taken as given by all players, i.e., to move first. For example, a

VER allows the exporter to "precommit" to not selling more than the level set by the VER. The VER is strategic in the sense that it alters the extent of competition among firms, and therefore can raise equilibrium profits of both foreign and domestic firms.

3. Recall that a game is described by a strategy space and a payoff function for each player. A strategy is a complete game plan, i.e., a rule telling the player what to do in *any* contingency. The strategic variable is an object that can be chosen by the firm. Thus, if price is the strategic variable for all firms, a firm's strategy is what price to choose in every contingency. A Nash equilibrium is a set of strategies for a game such that no player can choose a better strategy, given the Nash equilibrium strategies of the other players.

4. t could by *any* tax or subsidy. It could be on production, R&D, consumption, or trade, to mention a few examples. As usual, we refer to a tax on trade as tariff.

5. In fact the choice of strategic variable also affects whether a given ER has an effect or not. For example, if the ER is at the free trade level and strategic variable is quantity then an ER at the free trade level will have no effect while if the strategic variable is price, it will. If we consider market share restrictions, than an ER at the free trade level will have an effect whether the strategic variable is price or quantity.

6. Since much of the work on strategic trade policy has done so, we focus on studies which specify price or quantity as the strategic variable. When price (quantity) is the strategic variable, the model is that of Bertrand (Cournot) competition. Richer specifications are possible. For example, Kreps and Scheinkman (1983) show that for a particular two stage game, where capacity is chosen in the first stage and price is chosen in the second stage (with capacity as a constraint), the Cournot outcome occurs.

7. Recall that in a Stackelberg leader game, the foreign monopolist (leader) makes its choice before the potential entrant (follower). If the potential entrant enters the market, it chooses its output to maximize its profits, taking as given the output of the leader. The Stackelberg leader chooses its output level first, knowing that the follower will compete as a Cournot player if it enters.

8. The idea behind this approach is that the optimal policy to combat a distortion is that which affects the distortion at its source. For example, a labor market distortion is optimally targeted by appropriate intervention in the labor market, *not* a trade policy!

9. Other studies dealing with trade policy for imperfectly competitive agricultural institutions are Just, Schmitz, and Zilberman

(1979), Markusen (1984), and Thursby (1988). Rodrik (1989) does not explicitly look at agricultural institutions, but examines export policy for Indonesian nutmeg.

Our emphasis has been on policies when a single government is active. Of the agricultural studies, Thursby examines the policy equilibrium when both home and foreign governments are active. Brander and Spencer (1984b) and Dixit (1987a) also deal with the case where home and foreign governments are active.

10. This approach specifies a hypothetical parameter called conjectural variations which includes as special cases the Bertrand and Cournot models, and thus it is useful in explaining the role of the choice of strategic variable in determining optimal policy.

11. While they do not examine targeting or optimal policies and focus on a linear example, Markusen and Venables (1988) consider the welfare effects of policy with and without market segmentation.

12. This occurs for the same reason as in Brander and Spencer's (1981) rent extraction study with entry deterrence.

13. This non-equivalence of tariffs and quotas under monopoly is closely linked to that of the effects of ER.

14. This refers to the fact that the effects of a VER close to the free trade level are close to those of a VER at the free trade level.

15. Other threats are also possible. Threats are limited to being credible, i.e., it is required that they would actually be carried out.

16. Recent work by Driskill and McCafferty (1988) provides some basis for such a model using a dynamic markov game specification. This work is very preliminary, however, and has yet to be empirically implemented.

References

Abreu, D. 1984. *Infinitely Repeated Games with Discounting: A General Theory*. Harvard Institute of Economic Research Discussion Paper #1083.

———. 1985. *Extremal Equilibria of Oligopolistic Supergames*. Harvard Institute of Economic Research Discussion Paper #1167.

Bhagwati, J.N. 1965. On the Equivalence of Tariffs and Quotas. In *Trade, Growth and the Balance of Payments: Essays in Honor of Gottfried Haberer*, ed. R.E. Baldwin et al. Chicago: Rand McNally.

Brander J. and B. Spencer. 1981. Tariffs and the Extraction of Foreign Monopoly Rents Under Potential Entry. *Canadian Journal of Economics* 14:371-389.

―――. 1984a. Tariff Protection and Imperfect Competition. In *Monopolistic Competition and International Trade*, ed. Kierzkowski. Oxford: Blackwell.

―――. 1984b. Trade Warfare: Tariffs and Cartels. *Journal of International Economics* 16:227-242.

―――. 1985. Export Subsidies and Market Share Rivalry. *Journal of International Economics* 18:83-100.

Bull, C. 1986. Quota Induced Sales Games. New York University, mimeo.

Dixit, A. 1984. International Trade Policy for Oligopolistic Industries. *Economic Journal* 94 (supplement):1-16.

―――. 1986. Anti-dumping and Countervailing Duties with Oligopoly.

―――. 1987. Optimal Trade and Industrial Policy for the U.S. Automobile Industry. In *Empirical Methods for International Trade*, ed. Feenstra. Chicago: University of Chicago Press.

―――. 1987a. Strategic Aspects of Trade Policy. In *Frontiers of Economic Theory*, ed. T. Bewley. New York: Cambridge University Press.

Driskill R. and S. McCafferty. 1988. Dynamic Duopoly with Output Adjustment Costs in International Markets: Taking the Conjecture Out of Conjectural Variations. Ohio State University, mimeo.

Eaton, J. and G. Grossman. 1986. Optimal Trade and Industrial Policy Under Oligopoly. *Quarterly Journal of Economics* 100:383-406.

Grossman, G. and J.D. Richardson. 1985. *Strategic Trade Policy: A Survey of Issues and Early Analysis*. Special Papers in International Economics, no. 15, April, International Finance Section, Department of Economics, Princeton University.

Gruenspecht, H. 1988. Export Subsidies for Differentiated Products. *Journal of International Economics* 24:331-344.

Harris, R. 1985. Why Voluntary Export Restraints are Voluntary. *Canadian Journal of Economics* 18:799-809.

Helpman, E. and P. Krugman. 1989. *Trade Policy and Market Structure*. Cambridge: MIT Press.

Horstmann, I. and J. Markusen. 1986. Up the Average Cost Curve: Inefficient Entry and the New Protectionism. *Journal of International Economics* 20:225-248.

Itoh, M. and Y. Ono. 1982. Tariffs, Quotas, and Market Structure. *Quarterly Journal of Economics* 97:295-305.

―――. 1984. Tariffs vs. Quotas Under Duopoly of Heterogeneous Goods. *Journal of International Economics* 17:359-374.

Just, R., A. Schmitz, and D. Zilberman. 1979. Price Controls and Optimal Export Policies Under Alternative Market Structures. *American Economic Review* 69:706-15.

Katrak, H. 1977. Multinational Monopolies and Commercial Policy. *Oxford Economic Papers* 29:283-291.

Kreps, D. and J. Scheinkman. 1983. Quantity Precommitment and Bertrand Competition Yield Cournot Outcomes. *Bell Journal of Economics* 14:326-337.

Krishna, K. 1987. Tariffs vs. Quotas with Endogenous Quality. *Journal of International Economics* 23:97-122.

——. 1988. *Auction Quotas with Oligopoly*. National Bureau of Economic Research Working Paper no. 2723, September. Cambridge.

——. 1989a. *Auction Quotas with Monopoly*. National Bureau of Economic Research Working Paper no. 2840, February. Cambridge.

——. 1989b. Trade Restrictions as Facilitating Practices. *Journal of International Economics*, 26:251-270.

——. 1989c. What do VERs do? In *Beyond Trade Friction: Japan-U.S. Economic Relations*, ed. R. Sato and J. Nelson, 76-92. Cambridge: Cambridge University Press.

——. 1989. Protection and the Product Line. *International Economic Review*, forthcoming.

Krishna, K. and M. Itoh. 1987. Content Protection and Oligopolistic Interaction. *Review of Economics Studies* 55:107-125.

Krishna, K., K. Hogan and P. Swagel. 1989. The Non-optimality of Optimal Trade Policies: The U.S. Automobile Industry Revisited. Mimeo.

Krishna, K. and M. Thursby. 1988. *Optimal Policies with Strategic Distortions*. National Bureau of Economic Research Working Paper no. 2527. Cambridge.

Lambson, V. and J.D. Richardson. 1986. Tacit Collusion and Voluntary Restraint Arrangements in the U.S. Auto Market. Mimeo.

Markusen, J. 1984. The Welfare and Allocative Effects of Export Taxes versus Marketing Boards. *Journal of Development Economics* 14:19-36.

Markusen, J. and A. Venables. 1988. Trade Policy with Increasing Returns and Imperfect Competition: Contradictory Results from Competing Assumptions. *Journal of International Economics* 26:157-167.

Rodrik, D. 1989. Optimal Trade Taxes for a Large Country with Non-atomistic Firms. *Journal of International Economics* 26:157-167.

Rodrik, D. and C. Yoon. 1989. Strategic Trade Policy When Domestic Firms Compete Against Vertically Integrated Rivals. Mimeo.

Rotemberg, J. and G. Saloner. 1986. *Quotas and Stability of Implicit Collusion*. National Bureau of Economic Research Working Paper no. 1948. Cambridge.

Spencer, B. and R. Jones. *Vertical Foreclosure and International Trade Policy*. National Bureau of Economic Research Working Paper no. 2920. Cambridge.

Stegemann, K. 1989. Policy Rivalry Among Industrial States: What Can We Learn From Models of Strategic Trade Policy? *International Organization* 43:73-100.

Thursby, M. 1988. Strategic Models, Market Structure, and State Trading: An Application to Agriculture. In *Trade Policy Issues and Empirical Analysis*, ed. R. Baldwin. Chicago: University of Chicago Press.

Thursby, M. and J. Thursby. 1989. Strategic Trade Theory and Agricultural Markets: An Application to Canadian and U.S. Wheat Exports to Japan. Typescript.

Venables, A. 1985. *International Trade and Industrial Policy and Imperfect Competition: A Survey*. Discussion Paper no. 74, Center for Economic Policy Research, London.

Venables, A. and A. Smith. 1985. Trade and Industrial Policy Under Imperfect Competition. *Economic Policy* 3:621-672.

Discussion

Giovanni Anania

Kala Krishna and Marie Thursby have written an interesting and stimulating paper. They should be congratulated for the quality of their contribution. My comments as an agricultural economist focusing on international agricultural trade are those of an interested spectator, not an expert in government policy choice in imperfectly competitive markets. My discussion will be twofold. First, I will briefly review what I consider to be the main points outlined in the paper. Second, I will make a few comments and raise some questions.

The selective survey conducted by Kala and Marie is structured in four parts. Two broad groups of government interventions are considered ("indirect" policies, such as taxes or subsidies, and "direct" policies, such as VERs and quotas) and two imperfect market structures are taken into account (monopoly and oligopoly) for each of the two sets of policies.

In the first part of the survey, the well-known case when imports are supplied by a foreign monopolist and domestic demand is linear is discussed. In this scenario the government may extract foreign monopoly rent by imposing an import tax. The paper then discusses two other interesting cases. When a constant elasticity domestic demand is assumed, an import subsidy may actually improve welfare both at home and abroad. When fixed entry costs and Stackelberg behavior with the foreign firm acting as the leader are assumed, the imposition by the government of a sufficiently large import tariff may make it possible for a domestic firm to enter the market.

The second area of the literature reviewed in the paper addresses the case of a duopoly with Cournot competition and the domestic government maximizing domestic producers surplus. When this is the case the optimal policy is an export subsidy. The assumption made regarding the strategic variable is in this case crucial. If price instead of quantity competition is assumed, the government's optimal policy is an export tax.

In the case when a domestic monopoly coexists with a perfectly elastic competitive foreign supply, an import quota at the free trade level affects the market equilibrium by creating market power for the domestic firm when none existed. Domestic production and imports do not change, but domestic consumption decreases.

The last broad area of literature considered in the paper relates to the government intervening by imposing an import quota when an oligopolistic market structure occurs. If a duopoly with a Stackelberg structure where the domestic firm plays the role of the leader is assumed, then a quota at the free trade level increases profits both at home and abroad. The same result may be reached when the import quota is set at a level above the volume traded under free trade. When Stackelberg behavior is assumed, the hypothesis regarding the strategic variable is not an issue. However, the choice of the strategic variable becomes an issue when Nash competition is assumed. Under Cournot competition a quota at the free trade level leaves the market equilibrium unchanged. This is not the case when Bertrand competition is assumed. Profits increase both at home and abroad, and this may occur even with a quota set above the quantity imported under free trade.

There are two main general conclusions which I derive from the paper by Kala Krishna and Marie Thursby. The first conclusion is, quoting the authors, that "intuition one uses to guide policy makers with perfect competition does not apply, in general, to imperfectly competitive markets." The second one is that when policy decision making takes place in an imperfectly competitive market, assumptions regarding strategic variable, timing, entry and exit, and market segmentation become crucial. As the authors state, the literature on trade policies in imperfectly competitive markets presents a large number of models, based on different assumptions, and "often confusing and apparently conflicting set of associated results." Hence, the bottom line the paper seems to suggest is that, at least up to this point, the developments in the literature on government intervention with imperfect competition are far from providing us with a new paradigm.

Now, I will briefly address three specific points of the paper. The first point relates to the case when a domestic monopoly faces a perfectly elastic foreign supply and the domestic government intervenes with an import quota set at the free trade level. In the new equilibrium both the home and the foreign country import and export at the same time, with domestic production and imports unchanged, increased domestic price and decreased domestic consumption. In this case as well even a quota set above the free trade level may modify the market equilibrium creating monopoly power for the domestic firm. A point that is worth mentioning is that the imposition of a quota not only creates market power for the domestic firm, but, at the same time, it creates a quota rent. If import quotas are distributed free of charge, the market equilibrium is characterized by a two-price system for the foreign competitive producers. Those exporting to the domestic country receive a higher price than those selling on the foreign market (in the paper's Figure 2-4, the two prices equal 0H and 0A, respectively). If quotas are auctioned, the rent (AH times EB) is collected by the domestic government.

My second comment relates to the relevance of the assumption regarding the strategic variable when Nash competition occurs and the government optimal choice is restricted to "direct" policy instruments. This issue does not appear as a very strong barrier to the development of a theory leading to a more general set of results. When "direct" policies are used, the problem related to the choice of the strategic variable can by avoided by assuming that firms' reactions are based on the other firms' price and quantity signals. If the firms' reaction functions are based on profit maximizing behavior this approach seems more "natural" than the one assuming either price or quantity as the strategic variable.

Keeping the notation used in the paper, this means that the kth firm profit maximizing price and quantity reaction to its n-1 competitors is given by the solution to the following problem:

$$\text{Max}_{S_k} \; \Pi_k \, (S^1, S^2, ..., S^k, ..., S^n; V)$$

where the vector $S^i = (p^i, q^i)$ gives the ith firm's price and quantity reaction. Conceptually nothing is changed, but the problem is now cast in a higher dimension, and we can no longer use the familiar graphical approach to the duopoly case. Note that the information needed to derive each firm's reaction function does not differ from that needed when either price or quantity competition is assumed.

The third point relates to the framework the literature reviewed in the paper is cast in and to its relevance for empirical agricultural trade research. When a duopoly framework is considered, results are derived assuming either that there is no domestic consumption (both at home and in the foreign country) and that the domestic and the foreign firm are competing on a third market, or that they compete on the home market but that no consumption occurs in the foreign country. Both settings appear far from allowing the extension of the results reached to international agricultural market policy analysis.

In addition, all the literature reviewed explicitly assume that (a) only the domestic government intervenes (and moves first), and (b) firms only are active in the game (taking the government intervention as given). This is shown in Figure 2-6a, where G_H represents the home government, and F_H and F_F the home and the foreign firm, respectively. However, agricultural trade markets are characterized by all countries intervening and interacting with each other as well as with large firms (private marketing boards, for example) in the policy determination process. Such framework is represented in Figure 2-6b. The picture becomes even more complex if multinational trading companies (MTC in Figure 2-6c) holding market power are introduced as intermediaries between domestic and foreign firms and governments (in the form of public marketing boards). When this framework is considered the game structure clearly becomes complex and empirically hard to manage. A potentially fruitful and empirically approachable imperfect competition research framework for analyzing international agricultural markets, rather than assuming that firms only play the game, should include governments (both at home and abroad) and multinational trading firms intervening as active actors in the game (Figure 2-6d).

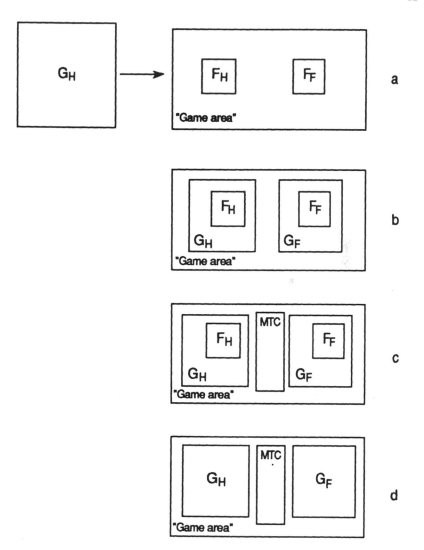

Figure 2-6. The Actors in the Game

3

International Trade, National Welfare, and the Workability of Competition: A Survey of Empirical Estimates

J. David Richardson

Introduction

The character, theory, and policy analysis of international trade have all changed markedly in the past ten years. Imperfectly competitive behavior seems increasingly relevant and perfect competition less. Technological advantage, scale economies, and multinational corporations seem to be playing growing roles in international trade. As a share of total trade and production for 14 large OECD countries, resource- and labor-intensive commodities have been shrinking steadily, and science-based, scale-intensive, and differentiated commodities and services have been growing; "intra-industry" trade has jumped dramatically in the 1980s after remaining constant during the 1970s (OECD 1987).

This paper surveys recent empirical research on the effects of international trade under imperfect competition. It is necessarily an *empirical* question whether or not an economy gains from trade in this environment. The easy presumption of gains from perfectly competitive mind-sets vanishes under imperfect competition. Yet as discussed in more detail below, empirical research has generated several replacement presumptions. The first is that as a rule, international trade still leads to gains, and when they are there, they are two to three times larger than those estimated under perfect competition.[1] The second and corollary presumption is that trade enhances the "workability" of competition.[2] The third, but much more tentative, presumption is that trade under imperfect competition *may* increase adjustment pressures on firms, workers, sectors, and nations precisely because competition is more workable.

Section II of the survey discusses the theoretical background for the empirical research. The algebra and graphics are admittedly stylized, and the examples discussed are decidedly hypothetical. The approach aims for clarity and accessibility, and its purpose is to distill a set of pure, unmixed elements that underlie the effects of trade under imperfect competition.[3]

The pure elements from Section II are joined in various combinations in two important types of recent empirical research surveyed there and in Section III: calibration and regression studies of the effects of trade on economic welfare, on market structure and performance, and on adjustment indicators.

Most of the recent empirical work assumes that manufacturing is the prime locale of imperfect competition. Section IV describes the few examples of empirical research on imperfectly competitive international agricultural markets,[4] with brief reference to an older but quite different style of research on imperfectly competitive trade in agricultural goods. Several future research directions are suggested.

This survey focuses on empirical research, not theory;[5] on recent contributions not surveyed elsewhere;[6] and on international economic issues rather than industrial organization more generally.[7]

Its most important conclusion is that simultaneous reduction of barriers to international and internal competition creates sizeable and mutually reinforcing increases in an economy's real income and market performance. There are exceptions, however. Such benefits are not virtually "guaranteed," in the way that they are in traditional textbook models of market economies with undistorted, perfect competition. Exceptions notwithstanding, the rule is that open trade *still* generates significant gains under imperfect competition.

Although there are sizeable estimated gains, these studies suggest a possibility that the blessings are not unmixed. Trade expansion may cause significant adjustment pressure—probably on firms and workers most heavily, but possibly also on entire industrial sectors and historically important trading partners. Recent calibration research does *not* support the blithe dismissal of adjustment pressure popular among those who emphasize specialization among mildly differentiated product lines and two-way intra-industry trade. Recent regression research, by contrast, finds little evidence that trade is any uniquely powerful source of forced exits of marginal firms, or of sharp stimuli for workers to move from sector to sector.

Theoretical Background

Both theory and empirical research on trade policy under imperfect competition have borrowed heavily from industrial organization. It is useful first to summarize some partial-equilibrium thinking about elementary industrial organization, and then to show how trade affects microeconomic market performance in the typical empirical study. Then the microeconomics is embedded in a general-equilibrium framework to show how trade affects national economic welfare in the typical empirical study.

Microeconomic Structure, Performance, and Trade

Most empirical studies of trade policy under imperfect competition use a very straightforward, yet very flexible, model of firm and industry behavior.[8] The model includes many realistic features, and also many familiar and robust economic relationships. For example, the familiar equality between marginal revenue and marginal cost implies a realistic kind of mark-up pricing, after some algebraic manipulation:

$$(p - c)/p = 1/e; \qquad (1)$$

where p and c are a product's price and marginal cost, and where e is the elasticity of demand that the firm perceives when it changes its price (defined positively).[9] Sensible firms will charge a mark-up over marginal cost (p - c), which when expressed as a proportion of price, is simply the reciprocal of the perceived demand elasticity. Elasticity governs market power. Perfect competitors facing infinitely elastic demand will enjoy no market power and no mark-up, but will be induced to price at exactly marginal cost (including of course the marginal cost of management, risk-bearing, and other entrepreneurial activity).

In imperfectly competitive settings, the first interesting question is how one firm's market power depends on the actions of its rivals. This can even be measured, and provides a first index of imperfect competition for empirical purposes. For example, suppose that n similar rival firms sell q units each of the same product in the same market. Then the total amount sold (nq) will in equilibrium be willingly purchased by buyers according to a market demand schedule:

$$nq = A - Bp, \qquad (2)$$

where A and B can be considered constants. This market demand schedule has its own elasticity E, which equals the reciprocal of $A/Bp - 1$.[10]

E, the market demand elasticity, will not in general be equal to e, each firm's perceived demand elasticity. It is helpful to see their relationship and the interdependence of each firm's market power along a continuum ordered by an "imperfection weight" w:

$$1/e = w(1/E). \tag{3}$$

At one extreme, for perfectly competitive firms, $w = 0$; imperfect competition plays no role, and firms are independent. At the other extreme, for a monopolist, $w = 1$, and e *is* E. For a tight collusion of n firms, acting as if they were one to maximize joint profits, w also $= 1$, and each firm faces an e that is equal to E. With less intensely collusive competition, w falls between 0 and 1, and each firm's market power depends moderately on that of its rivals. When w is empirically estimated (see Bresnahan 1987), it serves as one measure of the imperfection of competition.

A very important intermediate degree of imperfect competition is Cournot competition. It is a useful empirical reference point, in which w equals each firm's share of the overall market ($w = q/nq = 1/n$, and hence $e = nE$). Cournot competition is what emerges when each firm perceives as given the outputs of its rivals and then optimally decides on its own output.[11] "Cournot pricing," often encountered in empirical studies, is marking up price above marginal cost by the reciprocal of nE, the product of a firm's market share and the overall market elasticity.

The intensity of competition, measured by w, is one important dimension of imperfect competition. A second is profitability, connoting excess profits. Free entry and exit of firms drives excess profit rates per unit of output, r, close to zero in the long run.[12] In that case, the market structure is described as "monopolistically competitive." If n cannot vary, but is fixed by barriers to entry (or exit), then r is variable, and the market structure is called oligopolistic.

The excess profit rate r is defined more precisely as the proportion by which price lies above average cost per unit of product. Average cost is the sum of variable and fixed cost (f). Empirical studies often assume constant variable cost per unit, making

$$r = [p - c - (f/q)]/p. \tag{4}$$

When free entry and exit drive excess profits to zero, (4) implies that $(p-c)/p = f/pq$. In this case, a firm's mark-up over marginal cost from equation (1) is not arbitrary, but necessary to pay fixed cost per dollar of output. Market power is then merely the power to pay off one's fixed commitments to operate — legal incorporation and retainer fees, plant construction and maintenance, market research, licensing, and so on.

Often a finer distinction is made between "sunk" fixed costs, like initial incorporation and irrecoverable construction costs, and recoverable fixed costs, like retainer fees and plant maintenance. Sunk fixed costs are usually paid one time, and will be spread over however many periods that a product is produced. Recoverable fixed costs, often paid every period, are payments for inputs that are readily transferable to alternative uses, so that resources, though fixed, are never "wasted".[13]

Built into (4), and into the definition of average cost, is increasing returns to scale, in this case the ability to spread fixed costs thinner and thinner over larger and larger outputs. The sector described by equations (1) - (4) can be seen in fact as a type of natural monopoly. On the face of it, it would be wasteful for a duopoly to use up resources worth 2f when a monopoly would require only f to supply the whole market.

Equations (1) through (4) describe an imperfectly competitive industry producing a homogeneous product.[14] Price (p) and each firm's output (q) are almost invariably taken to be endogenous; cost conditions (c, f, and hence "scale elasticity" and "minimum efficient scale") and overall market demand parameters (A, B, and hence E) are almost invariably taken to be exogenous. Beyond those typical assignments, the behavioral system is very flexible. Either the number of firms (n) or each firm's excess profit rate (r) is selected as a third endogenous variable,[15] depending on whether free entry characterizes the industry or not. Either the perceived demand elasticity of each firm (e), or more rarely the imperfection weight (w), is taken as the fourth endogenous variable.[16]

Typical empirical research on the effects of international trade on market performance is based roughly on this microeconomic structure. Recent studies in this tradition are typically longitudinal regression studies, using panels of data across time and industries, and sometimes across firms themselves or countries. The dependent variable is typically some measure of endogenous performance, either static or dynamic, often price (p), the "price-cost margin" ($(p-c)/p$, or some monotonic transformation of it), entry and exit (changes in n over time), or "technical efficiency" (usually some transformation of average total cost, $c + (f/q)$, endogenous because of q, and reflecting efficiency as firms "rationalize," economizing on fixed costs by moving closer to "minimum efficient scale"). Several independent variables reflecting international trade are assumed, explicitly or implicitly, to proxy for the exogenous variables of the basic model above. A rise in import penetration ratios proxies for an exogenous increase in the number of rival firms, n, as if foreign firms penetrate progressively and symmetrically to their symmetric domestic rivals. A rise in export shares of each firm's sales, q, is a proxy for an addition to overall market demand (A-Bp) coupled with an increase in the number of rivals (n) in at least the export segment of the market.[17] An abolition of quotas is a proxy for a shift in overall market demand (A-Bp) coupled possibly with a decline in the ability to collude (decline in w).[18] Independent variables that are often taken to condition[19] the effects of import and export penetration are domestic concentration ratios (proxying for ex ante domestic n), organizational structure[20] (proxying perhaps for w), and multinationality (proxying perhaps for w, or for alternative outlets to merely the domestic A-Bp).[21] Conclusions from empirical research of this variety are discussed in Section III.

General Equilibrium Structure, Welfare, and Trade

Empirical studies of the effects of trade on national economic welfare under imperfect competition embed this microeconomics in a general-equilibrium framework, then isolate a number of important ways the imperfection influences the gains from trade. Three of the most important can be illustrated in a very simple diagram. Trade has potential to accentuate or alleviate an economy's welfare losses from: (1) distortionary pricing above marginal cost; (2) wasteful duplication of facilities or firms whose fixed costs cause a sector's average costs to be unduly high; (3) exploitative income transfers to foreign firms charging excess profits. After introducing the diagram, the case in which trade liberalization alleviates losses is discussed at length, followed by allusions to the remaining cases.

Figure 3-1 illustrates overall equilibrium for a hypothetical economy with one perfectly competitive sector, producing standardized goods (S), and a second imperfectly competitive sector, producing technology-intensive goods (T). The T sector will fit equations (1) - (4) above. Figure 3-1 can be taken initially to illustrate prohibitive trade barriers and a closed economy.[22]

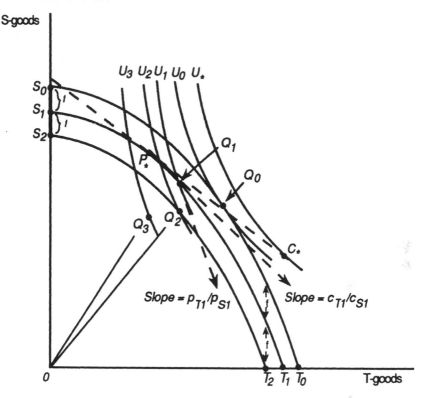

U = Equal-welfare contours

ST = Production-possiblities curves

Figure 3-1. A Stylized Economy Under Imperfect Competition

In order to produce even the first unit of T-goods, a fixed cost of f must be borne. Resources that could have produced S_0S_1 of standardized goods must be diverted, say, to a research laboratory for T. The economy's production possibilities curve $S_0S_1T_1$ lies uniformly inside of a reference curve that would pertain without fixed costs, S_0T_0.[23] Furthermore, if *two* firms compete by setting up research laboratories in order to produce T goods, the economy's production possibility curve would lie even lower: $S_0S_2T_2$. The second research laboratory may involve a social waste of resources equal to f, and the second firm's entry into the T market is possibly an example of inefficient entry.[24]

Since imperfectly competitive firms mark up price above marginal cost, equilibrium is illustrated in Figure 3-1 by a point like Q_1 for monopolistic market structure, and Q_2 for a duopoly. Buyers determine purchases at Q_1 so that their satisfaction from the last dollar's worth of each good bought is equal—illustrated by tangency between the relative price line p_{T1}/p_{S1} and the equal-welfare curve U_1. Imperfectly competitive mark-ups at Q_1 or Q_2 make the relative price of T goods higher than the relative marginal cost of T goods, c_T/c_S, which is what the slope of the production possibilities curve represents. The wedge between the two dashed lines at Q_1 represents a wasteful price distortion.

Finally, it is quite possible, for example at Q_2, that both firms are earning excess profits.[25] But both may be paying a portion of potentially larger excess profits to a foreign patent holder whose innovation the two research laboratories are implementing — a fixed fee, say, somewhat similar to the fixed costs f. In that case, there is a transfer of excess profits abroad, and the economy's real income, OQ_3, is less than its real output OQ_2.

Q_0 is a hypothetical reference point that locates the competitive equilibrium for this economy in the absence of any fixed costs. At least f of fixed costs is, however, an assumed fact of life, and the fundamental cause of imperfect competition. Thus the best the economy could hope to do is attain the equilibrium (undrawn) on $S_0S_1T_1$ that is tangent to an equal welfare contour like U_0, but below it and above U_1.

Relative to that "best" equilibrium, imperfect competition in this stylized economy can reduce welfare for three reasons. Price distortions can reduce welfare to U_1. Inefficient entry of a second T firm seeking excess profits can create unduly small-scale production and high average cost, reducing welfare further to U_2. And net payments of excess profits to imperfect competitors abroad can reduce welfare still further to U_3.

Now we can identify some extra potential gains from trade for an economy with imperfect competition. Liberalization that opens this particular economy to trade has all its normal benefits and more. Freer trade normally allows an economy to increase welfare to, say, U_*, by shifting production to a point like P_* and consumption to a point like C_*, with exports of S and imports of T respectively equal to the vertical and horizontal distances between P_* and C_*. But freer trade in this case also: (1) reduces imperfectly competitive price distortions, as every domestic firm is forced to compete against new foreign rivals; (2) "rationalizes" the domestic industry by forcing exit of excessive firms that drive up average costs; (3) reduces transfers of excess profits abroad. The economy's gains from freer trade, counting its effects on imperfect competition, are more like the difference between U_3 and U_* than between U_0 and U_*.

This accounting, however, is one-sided. It neglects to convey that most imperfectly competitive behavior is a two-edged sword. It can "cut" in *favor* of an economy as well as against it. Contrary to Figure 3-1, trade liberalization under imperfect competition is not guaranteed to produce extra benefits, either in theory or in practice. A simple alteration in the figure to make the economy an inherent exporter of T goods, instead of an importer, could show that: (1) mark-up pricing on imperfectly competitive exports can capture the same *benefits* as the classic optimal tariff under perfect competition; (2) having two dominant producers that have already sunk 2f of fixed costs in an export market (Boeing and McDonnell-Douglas) can deter undesirable entry by a foreign competitor (Airbus) that could potentially reduce the exporter's national welfare (see Krugman 1987); and (3) an economy's imperfectly competitive firms may on balance be *collectors* of excess profits on exports, which enhance its welfare. In this altered scenario, trade liberalization may reduce and even reverse the standard gains from trade. Trade liberalization may be detrimental to an economy, not beneficial, with imperfect competition.

Some of the elements in this fuller accounting, especially (3), are of course transfers from one economy to another. Thus from the viewpoint of all trading economies together, they are neither a gain nor a loss. Other elements, though, especially (1) and (2), apply at the global level as well: trade liberalization can be an effective instrument for disciplining distortionary forces and economizing on fixed resource costs—or, occasionally, it can accentuate distortions and resource costs.

We can draw an important conclusion about imperfectly competitive environments. From a national viewpoint, it is necessarily an *empirical* question whether there are gains from trade liberalization or losses, gains from active trade intervention or losses.

Figure 3-1 also illustrates the possibility of temporary short-run adjustment losses from trade liberalization. These turn out to be potentially larger under imperfect competition than under perfect competition. Trade penetration under imperfect competition due to scale economies can cause dramatic, discontinuous changes in trade, production, and market structure. Rationalization will usually imply that some plants or firms shut down, not just that they shrink. It may imply that a country loses *all* firms and production in a given sector.[26] For example, in Figure 3-1, a slight flattening of the dashed line P_* C_*, equivalent to small drop in world prices of technology-intensive goods, will cause the ideal production point to jump discontinuously from near P_* to S_0, without traversing intermediate points of incomplete specialization.[27] Both exports of S and imports of T would nearly double. Very little increase in welfare would result, but the industrial structure would be drastically altered. The T-industry would vanish! A very small, not very costly import barrier could then cause the industry to re-appear suddenly. That suddenness is precisely the point: trade and trade policy in some cases have very powerful effects on the sectoral composition of a country's production and employment under imperfect competition, without necessarily affecting its long-run welfare much.[28] But in the short run, obviously, welfare could decline if firms became suddenly insolvent, capacity became temporarily unproductive, and employees faced long dislocation and the need to move or retrain.

Several commentators summarize this adjustment concern and provide evidence.[29] Others, however, discount the concern. They suggest that what happens instead is that rationalization causes each country's plants or firms to specialize on narrowly defined *varieties* of a product, so that any dramatic changes in production and trade are of an "intra-industry" sort. A country may indeed cease producing large automobiles, but correspondingly increase its production and export of intermediate-sized models. Short-run adjustment costs will be minimal, they allege, because the same firms produce both varieties of auto, each of which uses very similar plants, machinery, workers, and techniques.[30]

Typical empirical research on the effects of international trade on national economic welfare is based roughly on this general-equilibrium structure. Recent studies in this tradition are typically "calibration/counterfactual" exercises.

Calibration/counterfactual research is in essence an empirical analog to comparative statics, and is familiar from computable-general-equilibrium (CGE) studies[31]—although applied here to industry studies as well, on the implicit assumption that the rest of the economy is like the S sector of Figure 3-1. The method begins with assumptions about economic behavior (such as equations (1) - (4) above), and maintains them as true for purposes of quantitative analysis. It then uses econo-metric estimates and industry case studies to measure key behavioral parameters. Since some parameters are subjective or have been estimated dubiously, there are always gaps. These can often be filled by assuming that the behavior accurately describes a real period, and using this period's data as a benchmark along with measured parame-ters to infer the values of missing, subjective, or dubious parameters. This inference is called "calibration," and amounts to making the assumed behavior and one period's data mutually consistent. The model's mechanics will consequently produce an equilibrium that matches reality for that one period. The counterfactual step is to change one (or more) of the parameters or data entries—in this case trade policy or penetration—and to calculate the new equilibrium that would have been generated by the model's mechanics. Values of vari-

ables in this new equilibrium are compared to their actual values—
"facts" are "countered" with hypothetical calculations—and differences between them are taken to be estimates of the effects of trade policy. The similarity to comparative statics should be clear.

Most of the studies of gains from trade in this tradition use the following procedure. Changes in trade penetration are taken to be due to some change in tariffs or other international differences in price (p), or to some change in the properties of the market demand curve (equation (2)), in the case of quotas, or to market-opening liberalization. Most studies rely on econometric estimates and industry data to measure the market demand behavior reflected in equation (2): average price, average quantity produced, market demand elasticity (E), etc.[32] Then the behavior summarized by equations (1) and (3) is "calibrated" in one of two ways. In the first, an assumption about inter-firm dependence (w) is made in (3), e.g., firms are collusive, or they are Cournot competitors, or. . . . Then the representative firm's perceived demand elasticity is inferred (i.e., e is inferred by (3) from an assumed w and an estimated E). Finally the inferred e and measured price are used in (1) to infer marginal cost (c), which is often not easy to measure. When marginal cost is measurable, however, usually from engineering or econometric studies, a second way of calibrating is often adopted. The measured c and measured p are used in (1) to infer e, the firm's perceived demand elasticity. It in turn, combined with estimates of E, implies a value for the intensity of competition, w, "calibrating" it instead of assuming it, using equation (3). Whichever method is used to establish c, e, and w, the values of marginal cost and prices can be used with equation (4): either to infer fixed costs, f, given data on excess profits r or the assumption that they are zero (free entry and exit); or to infer excess profits r, given engineering or econometric estimates of fixed costs, f. Occasionally, the value of a hard-to-measure trade policy is itself inferred using these techniques, as in the work of Baldwin and Krugman (1987, 1988).

Calibration/counterfactual methods have compelling strengths, despite their simplicity, selective and judgmental use of data and econometric estimation, insistence on maintaining rather than testing hypotheses, and imprecise statistical robustness (Baldwin 1988, Harrison et al. 1987). In the research surveyed here, they complement the data with a flexible structure to describe imperfect competition generically. They impose sensible economic consistency on experimentation (that is, incentives are calculated and profitable opportunities are assumed to be seized). And they organize the interpretation of results around accepted descriptions of economic trends (although there are usually several such descriptions). Not "anything can happen."

These strengths notwithstanding, calibration/counterfactual methods are more art than science.[33] They provide less definitive results than econometric, data-intensive methods that characterize modern empirical research in industrial organization. The intricacies and inadequacies of international and comparative national data for general-equilibrium application at the moment preclude recourse to more sophisticated empirical methods in the study of trade policy.

Conclusions from empirical research of this variety are discussed in Section III.

Empirical Conclusions

The most important conclusion from recent empirical research on these matters is that incorporating imperfectly competitive behavior, especially when motivated by scale economies, *can* make a significant difference to estimated effects of trade on economic welfare, industrial structure, and adjustment.

Table 3-1 summarizes conclusions from representative calibration studies discussed in Section II above.[34] The comparisons (small, moderate, large) are in every case to calibration research that assumes perfect competition and no fixed costs or scale economies. "Small" suggests little quantitative sensitivity to the inclusion of imperfect competition; "large" suggests considerable sensitivity.

Table 3-2 further documents the importance of imperfect competition for welfare calculations. It summarizes the results of several calibration studies capable of answering the question, "How would calculations have changed if fixed costs had been assumed to be zero and competition had been assumed perfect?"[35] In every case the calculations are estimates of the effect of various kinds of trade liberalization on the overall economic welfare of countries and regions. Economic welfare is defined as real income, a measure of the volume of goods and services that a given income can purchase, corresponding to the value of alternative U-curves in Figure 3-1.

Table 3-1. Recent, Representative Calibration Research on Trade Policy Under Imperfect Competition

Study	Size[1] of Effects on		
	Economic Welfare[2]	Market Performance[3]	Adjustment Stimuli[4]
Rodrik (1987)	moderate to large	moderate	moderate
Smith and Venables (1988a)	moderate	moderate	moderate to large
Digby, Smith, and Venables (1988)	moderate	moderate	moderate to large
Dixit (1988)	small	small	moderate
Baldwin and Krugman (1988)	?	large	large
Baldwin and Flam (1989)	small	moderate	moderate
Harris and Kwaka (1988)	?	moderate	small
Canada (1988a)	moderate	moderate	small
Brown and Stern (1988b)	small to moderate	small to moderate	small to moderate

[1] Approximate measure of responsiveness per "unit" of policy change (i.e., a rough elasticity). "Moderate" suggests responsiveness roughly twice as large as found in studies assuming perfect competition.
[2] Economic welfare effect of the policy change expressed as a percentage of the relevant sectoral or aggregate consumption.
[3] Effects on costs, price-cost margins, profits.
[4] Effects on a country's number and size of firms, output mix across sectors and/or patterns across trading partners.

Source: Modified and updated from Richardson (1989, Table 1).

Table 3-2. Welfare Effects[1] of Trade Policies Under Perfectly and Imperfectly Competitive[2] Assumptions (Percentage Change in Real Consumption)

Study/Experiment	Calculated Economic Welfare Impact Under Perfect Competition	Calculated Economic Welfare Impact Under Imperfect Competition	Effect on Calculation from Imperfect Competition[3]
Brown and Stern (1988a), Canada-U.S. free trade area.			
Canada	-0.015	1.177	1.192
U.S.	0.045	0.027	-0.018
Rest of World	-0.005	-0.004	0.001
Harris (1984), unilateral Canadian liberalization, reciprocated Canadian liberalization, effects on Canada.			
Unilateral	0.0	4.1	4.1
Reciprocated	2.4	8.6	6.2
Rodrik (1988),[4] 10 percent loosening of import quotas, effects on Turkey.			
No entry/exit			
Autos	6.3	2.6	-3.7
Tires	2.9	0.6	-2.3
Electrical appliances	1.0	-0.5	-1.5
Free entry/exit			
Autos	6.3	5.2	-1.1
Tires	2.9	4.1	1.2
Electrical appliances	1.0	1.2	0.2
Smith and Venables (1988a),[4] cut in transport/transfer costs among EC members equal to 2.5 percent of value of trade, effects on EC as a whole.			
No entry/exit			
Cement, lime, plaster	0.04	-0.10	-0.14
Pharmaceutical products	0.25	0.29	0.04
Artificial, synthetic fibers	0.91	0.99	0.08

(cont'd)

Table 3-2. (Continued)

Study/Experiment	Calculated Economic Welfare Impact Under Perfect Competition	Calculated Economic Welfare Impact Under Imperfect Competition	Effect on Calculation from Imperfect Competition[3]
Machine tools	0.56	0.84	0.28
Office machinery	0.59	0.88	0.29
Electric motors, generators	0.22	0.29	0.07
Electrical household appliances	0.49	0.64	0.14
Motor vehicles	0.62	0.83	0.21
Carpets, linoleum	0.47	0.67	0.20
Footwear	0.27	0.35	0.08
Free entry/exit			
Cement, lime, plaster	0.04	0.02	-0.02
Pharmaceutical products	0.25	0.29	0.04
Artificial, synthetic fibers	0.91	1.17	0.26
Machine tools	0.56	0.82	0.26
Office machinery	0.59	1.31	0.72
Electric motors, generators	0.22	0.29	0.07
Electrical household appliances	0.49	0.70	0.21
Motor vehicles	0.62	0.95	0.33
Carpets, linoleum	0.47	0.74	0.27
Footwear	0.27	0.37	0.10

[1]Calculated change in economic welfare as a percentage of GNP or GDP, except for Rodrik (1988) and Smith and Venables (1988a), where the calculated welfare effect is scaled by consumption within the industry indicated.

(cont'd)

Table 3-2. (Continued)

[2]Version reflected in table. Brown and Stern (1988a): monopolistic competition. Harris (1984): non-product differentiation. Rodrik (1988): Cournot pricing. Smith and Venables (1988a): Cournot pricing, models per firm constant.
[3]Second column minus first column.
[4]Column 1 estimates under perfect competition are especially rough approximations, by the author's own admission, but useful for an order of magnitude.

Underlying Sources
Brown and Stern (1988a, Table 3), scaled by 1976 base GDPs implied by Deardorff and Stern (1986, Table 4.4, pp.54-55): Canada--195,737; U.S.--1,737,250; Rest of World -- 3,020,124.
Harris (1984, Table 2, p. 1028).
Rodrik (1988, Tables 5-7).
Smith and Venables (1988a, Table 3, p. 1514).

Source
Richardson (1989, Table 2).

The principal conclusion from Table 3-2 is that on balance, trade liberalization has strong positive effects on economic welfare. These can be shown to be due in significant part to rationalization of industrial structure and heightened market competitiveness. Cases in which imperfectly competitive behavior shrinks or reverses the benefits from trade liberalization appear to be the exception rather than the rule, especially under the assumption of free entry to and exit from economic activity.

The representative regression studies of Table 3-3 and the older research along the same lines[36] support this conclusion. Trade liberalization, especially by encouraging import competition, disciplines the power of domestic firms to price above marginal cost, and provokes them toward both efficient scale and productivity enhancement.[37] Even the cases where no effects are detectable, especially those in Tybout's study of Chile, may be the end result of openness effectively leaving the economy free of imperfect competition "at the margin."

Table 3-3. Recent, Representative Regression Research on Market Performance Effects of Trade (Very rough estimated elasticities with respect to measures of import penetration)

Study/ Experiment	Price/ Cost Margin	Concentration Ratio	Technical Efficiency[a]	Entry/Exit Rates

Caves (1988) estimated effects of import and export growth on change in U.S. market performance between 1958 and 1977, controlling for domestic growth and selected measures of market structure; cross-section of changes between 1958 and 1977 in 65-70 selected four-digit manufacturing industries.

	N.A.	0.05	N.A.	None

de Ghellinick, Geroski, Jacquemin (1988); differences for Belgian manufacturing between estimates of market performance when completely open to and closed off from international trade; panel data on 82 three-digit industries annually 1973-78.

	-0.87[b]	N.A.	N.A.	N.A.

Morrison (1989); estimated effect of import prices on market performance in Canada; parallel time series of data for 7 manufacturing aggregates, 1960-82 (annually? quarterly?)

	-0.33[c]	N.A.	N.A.	N.A.

Roberts (1989); implicit effects (via import penetration) of increased Colombian trade restrictions 1981-84 after very slow liberalization 1976-80, with reliberalization 1985-86, but not back to 1980 levels; panel data on all[d] manufacturing plants annually 1977-85.

Average	-0.14	N.A.	0.04	None
Concentrated[e]	-.029	N.A.	0.23	None
Competitive[e]	-0.12	N.A.	None	None

Tybout (1989); implicit effects (via import penetration) of significant Chilean trade liberalization 1974-79 with mild retrenchment 1983-84; panel data on all[f] manufacturing plants annually 1979-85.

	None	N.A.	None	None

(cont'd)

60

Table 3-3. (Continued)

N.A. = Estimates were not attempted
None = No detectable influence

[a]Usually the rate of change over time in total factor productivity (Roberts 1989, Tybout 1989), sometimes relative to the United States (Baldwin and Gorecki 1986).

[b]de Ghellinick, Geroski, and Jacquemin (1988, p. 13) report estimates of the average price-cost margin across industries to be 0.069 if the Belgian economy were completely open, and 0.129 if it were completely closed. The percentage change from complete closure is therefore 100×[(0.129-0.069)/0.069]. The percentage change in import penetration from complete closure, starting from a base of m is 100×[(0-m)/m]. The ratio of these two percentage changes is -0.87/1.00, which is the elasticity recorded in the table.

[c]Unweighted average of seven industry estimates, also equal to estimate for total manufacturing aggregate; sign reversed for comparability with other entries in this table that are based on import penetration ratios assumed to vary negatively with import price.

[d]All plants with at least five employees in principle, except 1983-85, when most plants with less than 10 employees were excluded.

[e](Concentrated; Competitive) denote the elasticities for the (most; least) concentrated industries in Roberts' Colombian sample, measured by the Herfindahl index (H): (Petrol Derivatives, average H = 0.354; Food, average H = 0.008). For all Colombian industries during this period, average H = 0.090.

[f]Plants with more than 10 employees.

Version reflected in table

Caves (1988): Regression results with import penetration variable alone, without interactions between it and assets per employee, production worker share, or product-differentiation indicators.

Morrison (1989: See note c above.

Roberts (1989): Very provisional work in progress; intercept dummies for industry and year included as regressors.

Tybout (1989): Very provisional work in progress; intercept dummies for industry and year included as regressors.

Table 3-3. (Continued)

Sources

Caves (1988, Table 1.1, p. 14), coefficient multiplied by average ratio across all industries of variable DM to obtain rough elasticity.

de Ghellinick, Geroski, and Jacquemin (1988): See note b above.

Morrison (1989, Table 4).

Roberts (1989, Tables 6, 7, 9, and 12), using means across years of extreme values and means for all industries from Table 5 to obtain rough elasticities.

Tybout (1989, Table 5-8, 10).

The conclusions of Tables 3-1 and 3-3 are different, however, on the important question of adjustment pressures from trade. Recent regression studies find only small, often insignificant relationships between trade penetration and entry, exit, and concentration.[38] Yet recent calibration studies calculate moderate to severe pressures on workers to change industries and jobs, on firms to change outputs and activities, and on trading partners to change trade patterns.[39] As a corollary, the results in Table 3-2 suggest that the quantitative importance of imperfect competition is greatest when there is free entry and exit. That is, entry of new competitive firms, plants, and product lines, and exit of uncompetitive firms, plants, and product lines seem to create the largest change in counterfactual average resource productivity, and hence in economic welfare.[40] But these results are not paralleled strongly in the regression studies. So the severity of trade-induced adjustment pressures under imperfect competition remains empirically undetermined.

Agricultural Research and Potential

Most of the research above studies the influence of international trade in imperfectly competitive manufacturing sectors. Agricultural sectors are explicitly or implicitly assumed to have perfectly competitive market structure and trade. Both observation and a long research tradition,[41] however, maintain the opposite: imperfect competition is highly relevant to international trade in agricultural goods and to aspects of domestic marketing.

What sets this agricultural economics tradition apart from the research surveyed above is the frequency with which it invests governments and quasi-governmental agencies with market power. Marketing boards, purchasing agencies, and inter-governmental cartels are all featured prominently. Techniques from the study of imperfect competition—game theory, bargaining theory, differentiation of (national) products—are then applied to what are essentially variants of state trading.

This tradition has generated many useful insights, but not quite of the same focus as the calibration and regression research surveyed in this paper. It, by contrast, typically invests firms or plants with market power, not governments. It is more instructed by modern models of collusion and monopolistic competition than either the agricultural research or the older, industry-focused industrial-organization research. It is, in its regression-based variant, also more econometrically sophisticated. It has begun to employ potentially rich multi-dimensional panels of data (e.g. plants across firms across industries across countries across time), with corresponding formal attention to characteristics such as truncation, fixed effects, selectivity bias, and opportunities to exploit cross-equation constraints.

Among recent agricultural trade studies, Kolstad and Burris (1986), Lord (1989), and Mitchell and Duncan (1987) are traditional, with market power and product differentiation defined at the level of national aggregates: for wheat in Kolstad's and Burris' work; for ten crops or commodities (each independent of the others) in Lord's work; and for three crops (again independent) in Mitchell's and Duncan's work.

Anderson (1988) is less traditional, estimating the effects of U.S. cheese import quotas applied to an aggregate of interdependent commodities. Substitution and product differentiation play important roles in both supply and demand, but market power does not.

Thursbys' (1988, 1989 and ongoing) work comes closest to the distinctive features of recent research on manufactures.[42] It assigns potential market power to both firms and marketing boards in modelling international trade in wheat. In Thursby (1988), the focus is on optimal government taxes and subsidies on production, consumption, and trade.[43] She finds a configuration of "quasi-competitive" U.S. exporting firms and powerful-but-regulated marketing boards abroad to be consistent with time-series regressions that explain differentials between U.S. export and wholesale prices.[44] In Thursby and Thursby (1989), the focus is on competition for Japanese wheat purchases between the quasi-competitive U.S. exporting firms and the monolithic Canadian Wheat Board. A calibration approach, with calculated standard errors as a unique feature, suggests rivalry that is more competitive than Cournot assumptions would suggest.

In principle, empirical research on international trade and domestic adjustment for agricultural goods under imperfect competition would seem very promising. Governments *are* strategically active at the national level, as stressed traditionally; marketing boards and trading companies do raise interesting questions of agency and vertical integration; rapid technical progress creates dynamic scale economies and learning-by-doing effects; trade liberalization can radically alter important demand parameters by integrating near-perfectly segmented agricultural markets; "rationalization" of small-scale (family?) farms is the obvious cause of severe adjustment pressure on a well-defined segment of the population.

In practice, adequate data for this sort of research may be a problem. But the issues are extremely timely, with agricultural trade liberalization still high on the agenda of the Uruguay Round.

Notes

1.   "Gains" are measured by an economy's real income, its aggregate purchasing power over goods and services of all kinds.

2.   Frischtak et al. (1989) document this presumption in detail for 26 low and middle income economies with sizeable manufacturing sectors. Workable competition is an old term (attributed to J.M. Clark) with new relevance, describing ways that imperfectly competitive corporate and market structure can nevertheless generate outcomes that approximate perfect competition. The modern theory of contestable markets illustrates the idea.

3.   Section II expands and updates a similar section in Richardson (1989).

4.   One representative paper, by Thursby (1988, p. 80) observes that "the 'new' literature on the strategic use of trade policy under imperfect competition has . . . largely ignored the types . . . that can occur in agricultural markets," and a second, by Anderson (1988, p.13) comments that "empirical work on the ranking of simple tariffs and quotas in practice [with typical assumptions about competitive imperfections] for agriculture remains to be done."

5.   Sophisticated and detailed theoretical surveys exist in Dixit (1984), Grossman and Richardson (1985), Helpman (1984), Helpman and Krugman (1985, 1989), Krishna and Thursby (1989), Krugman (1985, 1986a,b), Markusen (1985), and Venables (1985).

6.   With varying attention to theoretical background, Baldwin and Gorecki (1985), Caves (1985), EC (1988, Ch. 6-7), Globerman (1988), Jacquemin (1982), and Kessides (1984) all survey empirical research on how international trade affects the competitive structure and performance of industries. Jacquemin summarizes an earlier unpublished

survey of the same material by Lyons (1979). Helpman and Krugman (1989, Ch. 9) and Richardson (1989) survey empirical research on how trade liberalization affects national economic welfare under imperfect competition, and Hazledine (1988, 1989) and Norman (1988) survey the methods and models underlying such research.

7.   Recent surveys of empirical research in industrial organization include Bresnahan (1987), Bresnahan and Schmalensee (1987), and Schmalensee (1988a,b).

8.   Rodrik (1988, Section 3) is a good example, quite parallel to the treatment here. See also Norman (1988).

9.   The elasticity of a firm's demand for units of its product, q, is the percent change in quantity demanded for every percent change in its price: $e = (\Delta q/q) + (\Delta p/p)$. Marginal revenue in this notation is defined as $\Delta(pq)$, which for small changes is approximately equal to $p(1-1/e)$. The mark-up expressed as a proportion of price is usually called the Lerner index of market power.

10. The elasticity of market demand, E, is the percent change in market quantity demanded for every percent change in market price: $E = (\Delta nq/nq) + (\Delta p/p)$, which $= (\Delta nq/\Delta p)(p/nq)$, which $= -B(p/A-Bp)$, which when defined positively $= 1/(A/Bp - 1)$.

11. If it is correct in its perceptions, then when it sells an extra unit it will force the market price received by itself and all other firms to decline by $1/B$. Hence it will perceive its own elasticity of demand, e, to be equal to $B*p/q$, which is exactly equal to nE (see note 10). Bresnahan (1987, pp.13, 74 passim) summarizes evidence in support of the view that the degree of competition associated with Cournot assumptions is empirically relevant, what ever one thinks of the rationality of the behavior. The unweighted average of estimated perceived elasticities (e) from his Table 1 is a little over 3 (using midpoints of intervals), higher than most estimated market demand elasticities (E), but well below the very large (infinite) estimates associated with perfect competition.

12. Zero may not be attained exactly if competition from the marginal entrant would make excess profits negative.

13. The distinction underlies the theory of contestable markets. Markets are said to be contestable if most fixed costs are recoverable, not sunk. Then although fixed costs may appear to be a barrier to entry and a source of the market power that allows excess profits to be earned, they are not. Entry can be engineered quickly and fluidly by any of a host of potential rivals, who need only transfer fixed-cost resources to the product in question without loss, and even a handful of large incumbents must price at exactly average cost to keep potential entry

from becoming actual entry. The distinction between sunk and recurrent fixed costs is also quite important for studying the dynamics of industrial structure, e.g., exactly when firms enter and exit an activity. But it has been less important in most early empirical research on trade under imperfect competition, which has had difficulty measuring the difference, and has also tended to focus on long-run equilibria. Krugman and Brainard (1988) is a first effort at calibrating the influence of trade employing a contestable-markets framework.

14. Product differentiation is discussed generally by Richardson (1989, p.16), and in agricultural applications by MacLaren (1989). In principle, product differentiation (in contrast to homogeneity) could be a fundamental cause of imperfect competition. This is most clearly seen in a Lancastrian "characteristics" approach to modelling fundamental demand behavior. Lancaster (1989) provides a recent application of his approach to the theoretical case for protection, and Lipsey (1987) describes the greater likelihood for adjustment pressures in such "characteristics" or "address" models. Very little empirical work on trade under imperfect competition adheres tightly to the Lancastrian approach, however. Levinsohn (1987) is the best example. Krishna, Hogan, and Swagel (1989) add rudimentary Lancastrian structure to Dixit's (1988a) calibration model, concluding that its calculations are quite sensitive to this addition. Lord exposits his regressions for agricultural export shares using Lancastrian language, but ultimately includes as regressors variables that are consistent with non-Lancastrian approaches as well.

15. Correspondingly, r is taken to be exogenous (equal to zero), or n.

16. Correspondingly, w is taken to be an exogenous structural measure of the degree of imperfection, or more rarely e, as in an asymmetric industry where each firm in the competitive fringe takes e to be exogenously infinite, unlike the dominant firms that are their rivals; see for an agricultural trade example, Mitchell and Duncan (1987).

17. It is easy to show that the effects of the export-dependence experiment are *theoretically* ambiguous, depending on how A, B, and n are shifted, whereas the import-penetration shock can be presumed almost always to vitiate market power. Empirical research in fact bears out both the ambiguity on one hand and the presumption on the other. See Caves (1985) and Kessides (1984) for summaries of research on the effects of trade dependence on price-cost margins. See Roberts (1989, pp.46-47) for results of the same flavor with respect to total factor productivity growth: it was positively correlated with growth of the domestic market and with import dependence (in concentrated industries), but not significantly correlated with export dependence.

18. See Krishna (1985).

19. In the typical regression studies, conditioning means that the coefficient on, for instance, import penetration, is [b + c(conditioning variable)], leading the researcher to insert both import penetration and its multiplicative interaction with the conditioning variable as regressors.

20. Roberts (1989) and Tybout (1989), for example, both examine differences across corporations, cooperatives, partnerships, and proprietorships.

21. Williamson (1986), Canada (1988, Ch.4).

22. Markusen (1985) provides a similar treatment.

23. $S_0S_1T_1$ is also no longer uniformly bowed out from the origin, given the S_0S_1 segment, creating the flavor of the non-convex production possibilities curves that are often associated with economies of scale.

24. The statement is merely illustrative. The possibility of *excessive* research and development is easily demonstrated under imperfectly competitive behavior. On the other hand, increased competition in producing research and development is often thought to increase its quantity and quality.

25. The ratio of average cost of T to S goods must lie between the slopes of the price line and the marginal cost line in this kind of model.

26. Whether it is firms, plants, or product lines that disappear depends on whether fixed costs (f) are associated with firms, plants, or product lines. The adjustment burdens are probably greatest for the first and least for the third, but only a little of the empirical research surveyed sheds light on this questions. Both Owen (1983) and Baldwin and Gorecki (1985, 1986) find that scale economies associated with plants seem more important for certain familiar measures of economic performance (e.g., bilateral trade balances, cost competitiveness) than those associated with firms and product lines. But their rich analyses also highlight many exceptions to this generalization, and do not specifically address the issue of adjustment. Ongoing research on panels of plants by Tybout (1987) and associates may address these adjustment issues more definitively. In results to date, both Roberts (1989) and Tybout (1989) have had trouble finding strong relationships between import penetration ratios and entry/exit rates of plants in Colombia and Chile, respectively. Roberts (1989, p. 59) finds, by contrast, a negative association between increased import penetration and output growth of *incumbent* plants.

27. The potential for sharper adjustment pressures is due to the reduced likelihood of diversified, non-specialized production in the

presence of fixed costs. The point can be seen in Figure 3-2, a re-drawing of Figure 3-1, and can be easily generalized to more realistic settings with many sectors. In the absence of fixed costs, the country's production remains diversified for all price ratios between m_0 and m_0'. When fixed costs are f, the country remains diversified for a much narrower band of price ratios, between m_1 and m_1'; when fixed costs are 2f, even narrower, between m_2 and m_2'.

28. This is what the theoretical literature implies when it concludes that trade patterns and the distribution of industries among trading partners is "indeterminate" under scale economies and imperfect competition (see Krugman (1985, pp.7-8, 23-24, 43), Helpman (1984, p.359)). This indeterminacy complicates a natural potential for empirical research on how imperfect competition may be a determinant of trade patterns, complementary to factor endowments in that regard.

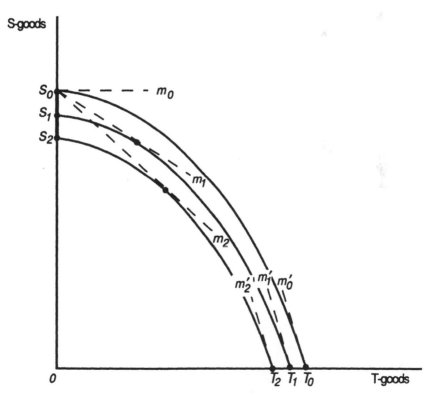

m = Relative-Price lines
ST = Production-possibilities curves

Figure 3-2. Imperfect Competition—A Stylized Example

The factor content of trade *is* determinate, however, under these assumptions. The factor content is the bundle of labor, capital, and other primary factor services embodied in exports and imports. This determinacy implies that long-run equilibrium differences among countries in factor regards will not be affected much by volatility in production and trade patterns caused by imperfect competition. But short-run dislocation and adjustment may nevertheless be frequent, burdensome, and welfare-reducing.

29. Harris (1985, pp. 165-166; 1986, pp.241-242), modified to account for normal turnover in Harris and Kwakwa (1988), Shea (1988), Wonnacott (1987, pp. 33-40), and Wonnacott with Hill (1987, Appendix B, C).

30. In Figure 3-1, if S and T were two varieties of a product with very similar production technologies, then the curves ST would be virtually straight lines. Moving resources from one corner to the other would be very easy, especially within the same firm. See also note 14 above on product differentiation.

31. Srinivasan and Whalley (1986) is the most relevant survey for trade policy. See also Borges (1986), Melo (1988), Shoven and Whalley (1984), and a large cliometric literature that uses the method. Burniaux et al. (1986) is a quite recent and synthetic example of CGE research, in this case applied to agriculture, and being expanded currently to encompass imperfectly competitive structure.

32. This is not as straightforward as it seems. There is an important technical issue to resolve at this step of the calibration. When (2) describes a national demand curve, then its cross-price elasticities with respect to similar products in other national markets range from zero (the case of "market segmentation") to finite values (characterized as the "Armington assumption," after one of its early developers). As such cross-price elasticities go to their limiting (infinitely large) values, however, then nationality of sales no longer differentiates a product, and (2) must define a global market. Econometric estimates can rarely be found in such a case. See Brown (1987), Brown and Stern (1988a), and Markusen and Venables (1988) for additional discussion.

33. Hence almost all the studies in this tradition perform elaborate sensitivity analysis with respect to key parameters. Some of these sensitivity analyses are multidimensional, e.g., in Markusen and Wigle (1988), and techniques for refining these are described by Wigle (1986) and by Bernheim, Scholz, and Shoven (1988).

34. Richardson (1989, Section III) provides a more detailed examination of most of these representative calibration studies, plus others that are similar to or derivative from them: Baldwin and

Krugman (1987, 1988), Condon and Melo (1986), summarized in Melo (1988, pp.481-484), Cox and Harris (1985, 1986), Cory and Horridge (1985), Daltung, Eskeland, and Norman (1987), Devarajan and Rodrik (1988), Dixit (1987c), Goto (1983, 1986, 1987), Gunasekera and Tyers (1988), Harris (1984, 1986), Harris with Cox (1984), Hazledine and Wigington (1987), Horridge (1987a,b), Klepper (1988), Krishna, Hogan, and Swaegel (1989), Kwakwa (1988), Laussel, Montet, and Peguin-Feissolle (1988), Lee (1988), Markusen and Wigle (1987, 1988), Melo and Tarr (1989, Ch. 7), Ngowsirimanee (1988), Nguyen and Wigle (1988), Owen (1983), Smith and Venables (1988a), Venables and Smith (1986, 1987), and Wigle (1988).

35. The comparisons are somewhat rough in several cases because perfectly competitive estimates were made in an admittedly crude way. This is especially true of Rodrik (1987), and Smith and Venables (1988a).

36. Recent regression studies similar to these include Baldwin and Gorecki (1986), Corbo, Tybout, and Melo (1988), Sleuwaegen and Yamawaki (1988), Williamson (1986), Yamawaki, Weiss, and Sleuwaegen (1989), and those being carried out for Morocco and Turkey in the fashion of Harrison (1989), Roberts (1989), and Tybout (1989), as part of the World Bank research project described in Tybout (1987). See the surveys in note 6 for references to earlier studies along these lines.

37. In encouraging exports also, the effects of trade liberalization on market performance can be rendered ambiguous, as discussed in note 17. But the estimated pro-competitive effects on the import side seem to resolve the ambiguity of the overall trade expansion in a pro-competitive direction.

38. Sleuwaegen and Yamawaki (1988) is an exception, finding larger (if insignificant) effects of trade penetration on concentration, and independent, significant, enormous (incredibly so) effects of intra-EC tariff abolition on concentration.

39. Harris and Kwakwa (1988) is an important exception, however. They find small adjustment pressures from trade once a normal process of entry, exit, and labor attrition is modeled and calibrated, too. Furthermore, Canada (1988b, pp. 32-35) finds very little difference between industrial and labor turnover for four trade-sensitive industries (leather, textiles, knitting mills, and clothing) and average turnover for all manufacturing. Finally, only a few of the studies in the tables, notably Smith and Venables (1988a) and Digby, Smith, and Venables (1988) incorporate product variety adequately enough to allow independent calculations of both inter-industry and intra-industry adjustment (they do not actually perform such a decomposition). Thus

the conclusion that calibration studies calculate significant adjustment pressures may be weakened by adequate modelling of variety in subsequent research.

40. Norman (1988) finds, however, that under free entry and exit the calculations summarized in Tables 3-1 and 3-2 are much more sensitive quantitatively to alternative parameter values and behavioral specifications than when there is a fixed number of firms.

41. See, for example, Karp and McCalla (1983), McCalla and Josling (1981), McCalla and Schmitz (1982), Paarlberg and Abbott (1984), and Schmitz, McCalla, Mitchell, and Carter (1981).

42. Caves and Pugel (1982) is a somewhat less recent counterpart.

43. In Thursby's many variants, marketing boards may or may not "represent" their firms, in the sense of maximizing joint producer profits. Governments enter independently of their own firms and marketing boards as a potential regulator of domestic price.

44. One problem with her regressions, however, and indeed with much similar work in agricultural trade, is that they have little power to rule out alternative explanations, specifically explanations that do not rest on imperfect competition. By contrast, the recent panel-based regression research in manufacturing seems less subject to the power-against-alternatives problem. A related problem is the absence of suitably rich data, forcing Thursby, for example, to employ a dummy variable to proxy for export subsidies, a time trend to proxy for distribution cost, and an interaction of the time trend and export volume to proxy for scale economies in distribution. Each maintained assumption that a proxy is accurate can be creatively denied by a critic intent on alternative explanations to imperfect competition for Thursby's regression results.

References

Anderson, James E. 1988. *The Relative Inefficiency of Quotas.* Cambridge: The MIT Press.

Aw, Bee Yan. 1989. *An Empirical Model of Mark-ups in a Quality- Differentiated Export Market.* Manuscript, Pennsylvania State University. July.

Baldwin, John R. and P. Gorecki. 1985. The Relationship Between Trade and Tariff Patterns and the Efficiency of the Canadian Manufacturing Sector in the 1970s. In *Canada-United States Free Trade,* by J. Whalley and R. Hill. Toronto: University of Toronto Press.

————. 1986. *The Role of Scale in Canada-US Productivity Differences.* Toronto: University of Toronto Press. Vol. 6. Royal Commission on the Economic Union and Development Prospects for Canada (the "Macdonald Commission").

————. *Structural Change and the Adjustment Process.* Economic Council of Canada, Ottawa, forthcoming.

Baldwin, Richard E. 1988. On Taking the Calibration Out of Calibration Studies. Partially completed manuscript, Columbia University, Graduate School of Business, July, work in progress.

Baldwin, Richard E. and Harvey Flam. 1989. Strategic Trade Policies in the Market for 30-40 Seat Commuter Aircraft. Institute for International Economic Studies Seminar Paper No. 431. Stockholm.

Baldwin, Richard E. and Paul Krugman. 1987. *Industrial Policy and International Competition in Wide-Bodied Jet Aircraft.* Manuscript in Baldwin (1988).

————. 1988. *Market Access and International Competition: A Simulation Study of 16K Random Access Memories.* in Feenstra (1988).

Baldwin, Robert E., ed. 1988. *Trade Policy Issues and Empirical Analysis.* Chicago: University of Chicago Press.

Bernheim, B. Douglas, J.K. Scholz, and J.B. Shoven. 1988. *Consumption Taxation in a General Equilibrium Model: How Reliable are Simulation Results.* Manuscript, Northwestern University.

Borges, Antonio M. 1986. Applied General Equilibrium Models: An Assessment of Their Usefulness for Policy Analysis. *OECD Economic Studies.* 7:7-43.

Bresnahan, Timothy F. 1987. *Empirical Studies of Industries with Market Power.* Stanford University, Center for Economic Policy Research, Publication No. 95, in Schmalensee and Willig (1988).

Bresnahan, Timothy F. and Richard Schmalensee. eds. 1987. *The Empirical Renaissance in Industrial Economics.* New York: Basil Blackwell.

Brown, Drusilla K. and R. M. Stern. 1987. A Modeling Perspective. in Stern, Trezise, and Whalley.

————. 1988. *Computational Analysis of the U.S.-Canada Free Trade Agreement: The Role of Product Differentiation and Market Structure.* Manuscript, presented at an April 29-30 Conference on Trade Policies for International Competitiveness, Cambridge, Massachusetts, sponsored by the National Bureau of Economic Research, forthcoming in Feenstra 1989.

————. 1988b. *Computable General-Equilibrium Estimates of the Gains from U.S.-Canadian Trade Liberalization.* Manuscript, May 20, presented at. a Lehigh University Conference on Economic Aspects of Regional Trading Arrangements, Bethlehem, Pennsylvania, May 25-27.

Burniaux, Jean-Marc, F. Delorme, I. Lienert, et al. 1988. *Quantifying the Economy-Wide Effects of Agricultural Policies: A General-Equilibrium Approach.* Organization for Economic Cooperation and Development, Department of Economics and Statistics, Working Paper No. 55.

Caves, Richard E. 1985. International Trade and Industrial Organization: Problems, Solved and Unsolved. *European Economic Review.* 28:377-395.

————. 1988. Trade Exposure and Changing Structures of U.S. Manufacturing Industries. in A. Michael Spence and H. A. Hazard, eds. 1988. *International Competitiveness.* Cambridge, Massachusetts: Ballinger.

Caves, Richard E. and T.A. Pugel. 1982. New Evidence on Competition in the Grain Trade. *Food Research Institute Studies.* 18:261-274.

Condon, T. and J. de Melo. 1986. Industrial Organization and Implications of QR Trade Regimes: Evidence and Welfare Costs. Manuscript, The World Bank.

Corbo, Vittorio, J. de Melo, and J. Tybout. 1988. *The Effects of Trade Policy on Scale and Technical Efficiency: New Evidence from Chile.* Manuscript, The World Bank.

Cory, Peter and M. Horridge. 1985. *A Harris-Style Miniature Version of ORANI.* IMPACT Research Centre Preliminary Working Paper No. OP-54, June.

Cox, David and R. G. Harris. 1985. Trade Liberalization and Industrial Organization: Some Estimates for Canada. *Journal of Political Economy.* 93:115-145.

————. 1986. A Quantitative Assessment of the Economic Impact on Canada of Sectoral Free Trade With the United States. *Canadian Journal of Economics.* 19:377-394.

Daltung, Sonja, G. Eskeland; and V. Norman. 1987. *Optimum Trade Policy Towards Imperfectly Competitive Industries: Two Norwegian Examples.* Centre for Economic Policy Research Discussion Paper No. 218, December.

Deardorff, Alan V. and R.M. Stern. 1986. *The Michigan Model of World Production and Trade.* Cambridge, Massachusetts: The MIT Press.

de Ghellinick, Elisabeth, P.A. Geroski, and A. Jacquemin. 1988. Inter-Industry Variations in the Effect of Trade on Industry Performance. *Journal of Industrial Economics.* 37:1-19.

de Melo, Jaime. 1988. Computable General Equilibrium Models for Trade Policy Analysis in Developing Countries: A Survey. *Journal of Policy Modeling.* 10:469-503.

de Melo, Jaime and D. Tarr. 1989. *A General Equilibrium Analysis of U.S. Trade Policy.* Manuscript, The World Bank, April 5.

Department of Finance, Fiscal Policy and Economic Analysis Branch. 1988a. *The Canada-U.S. Free Trade Agreement: An Economic Assessment.* Ottawa, Canada.

Devarajan, Shantayanan and D. Rodrik. 1988. Trade Liberalization in Developing Countries: Do Imperfect Competition and Scale Economies Matter? *American Economic Review.* 79:283-287.

Digby, Caroline, A. Smith. and A. Venables. 1988. *Counting the Cost of Voluntary Export Restrictions in the European Car Market.* Centre for Economic Policy Research Discussion Paper No. 249, June.

Dixit, Avinash. 1984. International Trade Policy for Oligopolistic Industries. *Economic Journal.* 94 (supplement), pp. 1-16.

———. 1987. Tariffs and Subsidies Under Oligopoly: The Case of the U.S. Automobile Industry. in H. Kierzkowski, ed., *Protection and Competition in International Trade.* Oxford: Basil Blackwell.

———. 1988. Optimal Trade and Industrial Policies for the US Automobile Industry. in Feenstra 1988.

EC (European Communities), Commission of 1988. The Economics of 1992. *European Economy.* No. 35 (March).

Economic Council. 1988b. *Adjustment Policies for Trade-Sensitive Industries.* Ottawa, Canada.

Eichengreen, Barry and L.H. Goulder. 1988. *A Computable General Equilibrium Model for Analyzing Dynamic Responses to Trade Policy and Foreign Competition.* Final Report to the U.S. Department of Labor, Bureau of International Labor Affairs, Office of Foreign Economic Research, March.

Feenstra, Robert C., ed. 1988. *Empirical Methods for International Trade.* Cambridge, Massachusetts: The MIT Press.

———. 1989. *Exchange Rate and Trade Policies for International Competitiveness.* Chicago: University of Chicago Press, forthcoming.

Frischtak, Claudio, U. Zachau, and B. Hadjimichael. 1989. *Competition Policies for Industrializing Countries.* World Bank; Industry Development Division; Industry and Energy Department; Policy, Planning, and Research. Manuscript, March 20.

Globerman, Steven. 1988. *The Impacts of Trade Liberalization on Imperfectly Competitive Industries: A Review of Theory and Evidence.* Economic Council of Canada Discussion Paper No. 341, January.

Goto, Junichi. 1985. *A General Equilibrium Analysis of Trade Restrictions Under Imperfect Competition: Theory and Some Evidence for the Automotive Trade.* World Bank Development and Research Department Discussion Paper DRD #130.

———. 1986. *A General Equilibrium Analysis of International Trade Under Imperfect Competition in Both Product and Labor Market -- Theory and Evidence from the Automobile Trade.* Unpublished doctoral dissertation, Yale University, May.

———. 1987. *International Trade and Imperfect Competition—Theory and Application to the Automobile Trade.* Manuscript, The World Bank, October.

Grossman, Gene M. 1989. *Promoting New Industrial Activities: A Survey of Recent Arguments and Evidence.* Manuscript, Organization for Economic Cooperation and Development, August.

Grossman, Gene M. and E. Helpman. 1988. *Product Development and International Trade.* National Bureau of Economic Research Working Paper No. 2540, May.

———. 1989a. *Comparative Advantage and Long-Run Growth.* National Bureau of Economic Research Working Paper No. 2809, January.

———. 1989b. *Endogenous Product Cycles.* National Bureau of Economic Research Working Paper No. 2913, March.

———. 1989c. *Growth and Welfare in a Small Open Economy.* National Bureau of Economic Research Working Paper No. 2970, July, revised.

Grossman, Gene M. and J. D. Richardson. 1985. Strategic U.S. Trade Policy: A Survey of Issues and Early Analysis. Princeton University *Special Papers in International Economics.* No. 15.

Gunasekera, H. Don B. H. and R. Tyers. 1988. *Imperfect Competition and Returns to Scale in a Newly Industrializing Economy: A General Equilibrium Analysis of Korean Trade Policy.* Presented at the 1988 Australian Economics Congress, Australian National University, Canberra, August 28-September 2.

Harris, Richard G. 1984. Applied General Equilibrium Analysis of Small Open Economies with Scale Economies and Imperfect Competition. *American Economic Review.* 74:1016-1032.

———. 1985. *Summary of a Project on the General Equilibrium Evaluation of Canadian Trade Policy.* in Whalley and with Hill.

———. 1986. *Market Structure and Trade Liberalization: A General Equilibrium Assessment.* in Srinivasan and Whalley.

————. 1988. *A Guide to the GET [General Equilibrium Trade] Model*. Working Paper, Department of Finance, Canada.

Harris, Richard G. and V. Kwakwa. 1988. *The 1988 Canada-United States Free Trade Agreement: A Dynamic General Equilibrium Evaluation of the Transition Effects*. Presented at a July 8-9 Conference on Strategic Trade Policy, University of Sussex, England, sponsored by the National Bureau of Economic Research and the Centre for Economic Policy Research.

Harris, Richard G. with David Cox. 1984. *Trade, Industrial Policy and Canadian Manufacturing*. Toronto: Ontario Economic Council.

Harrison, Ann E. 1988. *Exchange-Rate Pass-through and Imperfect Competition*. Manuscript, Princeton University, Department of Economics, March.

————. 1989. *Productivity, Imperfect Competition and Trade Liberalization in the Cote d'Ivoire*. Manuscript, The World Bank, June 9.

Harrison, Glenn W., R. Jones, L.J. Kimbell, and R. Wigle. 1987. *How Robust is Applied General Equilibrium Analysis?* University of Western Ontario, Centre for the Study of International Economic Relations, Working Paper No. 8707C, May.

Hazledine, Tim. 1988. Industrial Organization Foundations of Trade Policy: Modeling The Case of Canada - U.S. Free Trade. Presented at the Fifteenth Annual Conference of the European Association for Research in Industrial Economics, August 31-September 2, forthcoming *Australian Journal of Agricultural Economics*.

————. 1989. *Why the Free-trade Gain Numbers Differ So Much: Analysis of an encompassing general equilibrium model*. Manuscript, Department of Agricultural Economics, University of British Columbia, May 17.

Hazledine, Tim and I. Wigington. 1987. Protection in the Canadian Automobile Market: Costs, Benefits, and Implications for Industrial Structure and Adjustment. in OECD, *The Costs of Restricting Imports: The Automobile Industry*, Paris.

Helpman, Elhanan. 1984. Increasing Returns, Imperfect Markets, and Trade Theory. in R.W. Jones and P.B. Kenen, eds., *Handbook of International Economics*. Amsterdam: North-Holland.

Helpman, Elhanan and P.R. Krugman. 1985. *Market Structure and Foreign Trade*. Cambridge, Massachusetts: The MIT Press.

————. 1989. *Trade Policy and Market Structure*. Cambridge, Massachusetts:The MIT Press.

Horridge, Mark. 1987a. *Increasing Returns to Scale and the Long Run Effects of a Tariff Reform*. IMPACT Research Centre Preliminary Working Paper No. OP-62, August.

————. 1987b. *The Longterm Costs of Protection: An Australian Computable General Equilibrium Model*. unpublished doctoral dissertation, University of Melbourne.

Jacquemin, Alexis. 1982. Imperfect Market Structure and International Trade: Some Recent Research. *Kyklos.* 35:75-93.

Karp, L.S. and A.F. McCalla. 1983. Dynamic Games and International Trade: An Application to the World Corn Market. *American Journal of Agricultural Economics.* 65:641-656.

Kessides, Ioannis. 1984. *Industrial Organization and International Trade: Some Recent Developments.* World Bank Country Policy Department, CP Discussion Paper No. 1984-32, June.

Klepper, Gernot. 1988. *Simulating Competition in the Market for Large Transport Aircraft.* Presented at a July 8-9 Conference on Strategic Trade Policy, University of Sussex, England, sponsored by the National Bureau of Economic Research and the Centre for Economic Policy Research.

Kolstad, C.E. and A.E. Burris. 1986. Imperfectly Competitive Equilibria in International Commodity Markets. *American Journal of Agricultural Economics.* 68:25-36.

Krishna, Kala. 1985. *Trade Restrictions as Facilitating Practices.* National Bureau of Economic Research Working Paper No. 1546, January.

Krishna, Kala, K. Hogan, and P. Swagel. 1989. *The Non-Optimality of Optimal Trade Policies: The U.S. Automobile Industry Revisited, 1979-1985.* Manuscript, Harvard University, Department of Economics, March.

Krishna, Kala and M. Thursby. 1989. *Trade Policy with Imperfect Competition: A Selective Survey.* This volume.

Krugman, Paul R. 1985. *Increasing Returns and the Theory of International Trade.* National Bureau of Economic Research Working Paper No. 1752.

———. 1986a. *Industrial Organization and International Trade.* National Bureau of Economic Research Working Paper No. 1957, June, in Schmalensee and Willig (1988).

———. ed. 1986b. *Strategic Trade Policy and the New International Economics.* Cambridge, Massachusetts: The MIT Press.

———. 1987. Is Free Trade Passe? *Journal of Economic Perspectives.* 1:131-144.

Krugman, Paul R. and L. Brainard. 1988. *Problems in Modelling Competition in the Aircraft Industry.* Presented at a July 8-9 Conference on Strategic Trade Policy, University of Sussex, England, sponsored by the National Bureau of Economic Research and the Centre for Economic Policy Research.

Kwakwa, Victoria. 1988. *Sequential General Equilibrium Analysis of Canadian Trade Policy.* Unpublished doctoral dissertation, Queen's University, Kingston, Ontario.

Lancaster, Kelvin. 1984. Protection and Product Differentiation. in Henry Kierzkowski, ed., *Monopolistic Competition and International Trade.* Oxford: Oxford University Press.

————. 1989. *The 'Product Variety' Case for Protection*. Columbia University Department of Economics Discussion Paper Series No. 423.

Laussel, Didier, C. Montet, and A. Peguin-Feissolle. 1988. Optimal Trade Policy Under Oligopoly: A Calibrated Model of the Europe-Japan Rivalry in the EEC Car Market. *European Economic Review*. 32:1547-1565.

Lee, Hiro. 1988. *Imperfect Competition, Industrial Policy, and Japanese International Competitiveness*. Unpublished doctoral dissertation, University of California, Berkeley, September.

Levinsohn, James. 1987. *Empirics of Taxes on Differentiated Products: The Case of Tariffs in the U.S. Automobile Industry*. in Baldwin (1988).

Levinsohn, James and R. Feenstra. 1988. *Identifying the Competition*. Manuscript, University of Michigan, Department of Economics, July.

Lipsey, Richard G. 1987. Models Matter When Discussing Competitiveness: A Technical Note. in Richard G. Lipsey and Wendy Dobson, eds., *Shaping Comparative Advantage*. Scarborough, Ontario: Prentice-Hall Canada, for the C.D. Howe Institute, Policy Study No. 2.

Lord, Montague J. 1989. Product Differentiation in International Commodity Trade. *Oxford Bulletin of Economics and Statistics*. 51:35-53.

Lyons, B. 1979. *International Trade, Industrial Pricing and Profitability: A Survey*. Presented at the 6th Conference of EARIE, Paris, manuscript.

MacLaren, Donald. 1989. *Implications of New Theory for Modelling Imperfect Substitutes in Agricultural Trade*. this volume.

Markusen, James R. 1985. *Canadian Gains from Trade in the Presence of Scale Economies and Imperfect Competition*. In Whalley with Hill (1985).

Markusen, James R. and A.J. Venables. 1988. Trade Policy with Increasing Returns and Imperfect Competition: Contradictory Results from Competing Assumptions. *Journal of International Economics*. 24:299-316.

Markusen, James R. and Randall Wigle. 1987. *U.S.-Canada Free Trade: Effects on Welfare and Sectoral Output/Employment in the Short and Long Run*. Research report to the U.S. Department of Labor, Bureau of International Labor Affairs, Office of Foreign Economic Research.

————. 1988. Nash Equilibrium Tariffs for the U.S. and Canada: The Roles of Country Size, Scale Economies, and Capital Mobility. Manuscript, Wilfred Laurier University, Department of Economics, February, forthcoming, *Journal of Political Economy*.

McCalla, Alex F. and T.E. Josling. 1981. *Imperfect Markets in Agricultural Trade*. Montclair, New Jersey: Allanheld, Osmum.

McCalla, Alex F. and A. Schmitz. 1982. State Trading in Grain. in M.M. Kostecki, ed., *State Trading in International Markets*. New York: St. Martin's Press.

Mitchell, Donald O. and R.C. Duncan. 1987. Market Behavior of Grains Exporters. World Bank *Research Observer*. 2:3-21.

Morrison, Catherine J. 1989. *Markup Behavior in Durable and Nondurable Manufacturing: A Production Theory Approach.* National Bureau of Economic Research Working Paper No. 2941, April.

Ngowsirimanee, Teerana. 1988. *Monopolistic Competition and Trade Liberalization in a Small Open Developing Country: A Computable General Equilibrium Analysis.* Unpublished doctoral dissertation, University of Wisconsin-Madison, September.

Nguyen, Trien T. and R.M. Wigle. 1988. *Trade Liberalization With Imperfect Competition: The Large and Small of It.* Manuscript, University of Waterloo, Department of Economics, March.

Nguyen, Trien T., J. Whalley, and R.M. Wigle. 1988. *Three Variants of the Whalley Model of Global Trade.* Manuscript, University of Waterloo, Department of Economics, September.

Norman, Victor D. 1988. *Trade Under Imperfect Competition—Theoretical Ambiguities and Empirical Irregularities.* Presented at the European Economic Association Annual Meetings, Bologna, August.

Norton, R.D. 1986. Industrial Policy and American Renewal. *Journal of Economic Literature.* 24:1-40.

OECD (Organization for Economic Cooperation and Development). 1987. *Structural Adjustment and Economic Performance.* Paris.

Owen, Nicholas. 1983. *Economies of Scale, Competitiveness, and Trade Patterns Within the European Community.* Oxford: Oxford University Press.

Paarlberg, Philip L. and P.C. Abbott. 1984. Toward a Countervailing Power Theory of World Wheat Trade. in G.G. Storey, A. Schmitz, and A.H. Sarris, eds., *International Agricultural Trade.* London: Westview Press.

Richardson, J. David. 1989. Empirical Research on Trade Liberalization with Imperfect Competition: A Survey. *OECD Economic Studies.* 12:7-50.

Roberts, Mark J. 1989. *The Structure of Production in Colombian Manufacturing Industries.* Manuscript, The World Bank, May 2.

Rodrik, Dani. 1988. *Imperfect Competition, Scale Economies, and Trade Policy in Developing Countries.* in Baldwin (1988).

Schmalensee, Richard. 1988a. Industrial Economics: An Overview. *The Economic Journal.* 98:643-681.

———. 1988b. *Empirical Studies of Rivalrous Behavior.* Associazione Borsisti Luciano Jona, Working Paper No. 18, May.

Schmalensee, Richard and R. Willig, eds. 1988. *Handbook of Industrial Organization.* Amsterdam: North-Holland.

Schmitz, Andrew. A.F. McCalla, D.O. Mitchell and C. Carter. 1981. *Grain Export Cartels.* Cambridge, Massachusetts: Ballinger.

Shea, Brian F. 1988. *The Canada-United States Free Trade Agreement: A Summary of Empirical Studies and An Industrial Profile of the Tariff Reductions.* U.S. Department of Labor, Bureau of International Labor

Affairs, Office of Foreign Economic Research, Economic Discussion Paper No.28, March.

Shoven, John B. and J. Whalley. 1984. Applied General-Equilibrium Models of Taxation and International Trade. *Journal of Economic Literature.* 22:1007-1051.

Sleuwaegen, Leo and Hideki Yamawaki. 1988. The Formation of the European Common Market and Changes in Market Structure and Performance. *European Economic Review.* 32:1051-1525.

Smith, Alisdair and A. Venables. 1988a. Completing the Internal Market in the European Community: Some Industry Simulations. *European Economic Review.* 32:1501-1525.

―――. 1988b. The Costs of Non-Europe: An Assessment Based on a Formal Model of Imperfect Competition and Economies of Scale. In European Community, Studies on the Economics of Integration, Volume 2 of *Research on the "Cost of Non-Europe": Basic Findings.* Brussels: European Community Publications Office.

Srinivasan, T.N. and J.Whalley, eds. 1986. *General Equilibrium Trade Policy Modeling.* Cambridge, Massachusetts: The MIT Press.

Tarr, David G. 1989. *A General Equilibrium Analysis of the Welfare and Employment Effects of U.S. Quotas in Textiles, Autos, and Steel.* U.S. Federal Trade Commission, Bureau of Economics Staff Report, Washington, February.

Thursby, Marie. 1988. *Strategic Models, Market Structure, and State Trading: An Application to Agriculture.* in Baldwin (1988).

Thursby, Marie C. and J.G. Thursby. 1989. *Strategic Trade Theory and Agricultural Markets: An Application to Canadian and U.S. Wheat Exports to Japan.* this volume.

Tybout, James. 1987. *Industrial Competition, Productive Efficiency, and Their Relation to Trade Regime: Project Narrative.* Manuscript, The World Bank, May 3.

―――. 1989. *Entry, Exit, Competition and Productivity in the Chilean Industrial Sector.* Appendix by Lili Liu, manuscript, The World Bank, May 3.

Venables, Anthony J. 1985. International Trade, Trade and Industrial Policy and Imperfect Competition: A Survey. Centre for Economic Policy Research Discussion Paper No. 74.

Venables, Anthony and A. Smith. 1986. Trade and Industrial Policy Under Imperfect Competition. *Economic Policy.* 3:621-672.

―――. 1987. *Trade and Industrial Policy Under Imperfect Competition: Some Simulations for EEC Manufacturing.* Manuscript, presented at a September 17 Conference on Empirical Studies of Strategic Trade Policy, Cambridge, Massachusetts, sponsored by the National Bureau of Economic Research and the Centre for Economic Policy Research.

Whalley, John with R. Hill. 1985. *Canada-United States Free Trade.* Toronto: University of Toronto Press. Volume 11 in the research program of the Royal Commission on the Economic Union and Development Prospects for Canada (the "Macdonald Commission").

Wigle, Randall. 1986. *Numerical Modeling of Global Trade Issues: Facing the Challenge from Punta del Este.* Manuscript, presented at a November 7 Workshop on Modeling and Analytical Issues in the New GATT Round, Washington, D.C., sponsored by the University of Western Ontario and University of Michigan.

———. 1988. Canadian Trade Liberalization: Economies of Scale in a Global Context. *Canadian Journal of Economics.* 21:539-564.

Williamson, Peter J. 1986. Multinational Enterprise Behavior and Domestic Industry Adjustment Under Import Threat. *Review of Economics and Statistics.* 68:359-368.

Wonnacott, Paul. 1987. *The United States and Canada: The Quest for Free Trade.* Washington: Institute for International Economics Policy Analyses in International Economics No.16, March.

Wonnacott, Ronald J. with R. Hill. 1987. *Canadian and U.S. Adjustment Policies in a Bilateral Trade Agreement.* Toronto and Washington, D.C.: Canadian-American Committee of the C.D. Howe Institute (Canada) and National Planning Association (U.S.A.).

Yamawaki, Hideki, L.W. Weiss, and L. Sleuwaegen. 1989. Industry Competition and the Formation of the European Common Market. in Leonard W. Weiss, ed. *Concentration and Price.* Cambridge, Massachusetts: The MIT Press, forthcoming.

Discussion

Philip L. Paarlberg

This paper by J. David Richardson is an excellent overview of the empirical applications of imperfect competition theory as applied in economics. It is reassuring that the gains from trade are retained in imperfect markets and may even be magnified.

As noted in the paper the empirical analyses are largely hypothetical. This raises the concern that the empirical conclusions presented may not be as robust as suggested in the paper. To illustrate the sensitivity of imperfect competition models, this discussion examines several characteristics of the behavior of international agricultural (grain) trading firms. The implications for empirical work on these traders are highlighted to show how researcher decisions would influence the magnitudes obtained in empirical analyses.

The Production Function

Normally firms are assumed to consume inputs and transform them into an output. International agricultural trading firms perform a different role. These middlemen transform grain spatially; that is, they move the grain from farmers to foreign customers. They may also transform the product temporally. Whether or not these firms change the grain's physical characteristics is a subject of debate. Critics of these traders argue that the middlemen lower the grain's quality as they move it through the marketing channel.

How inputs and outputs are defined may effect the empirical estimates. One framework for analyzing middlemen production might borrow from the "new household economics" literature where demand and supplies for characteristics are used (Becker; Lancaster). This could possibly introduce product differentiation whereas a traditional view of the production process would be that the grain is homogeneous. For example, U.S. number 2 corn sold by Cargill might differ from that sold by Continental due to differences in product servicing or contract specification. Such changes in interpreting the production process might lead to alternative predictions of middlemen behavior following a policy shock.

Payoff Functions and Rival Behavior

Because most international agricultural trading firms are privately owned there is little known about their payoff functions and their interaction with rivals. Most analyses assume profit maximization as the payoff functions and structure the game as Cournot oligopoly. It is not certain that these assumptions are correct for the grain traders.

The choice of payoff functions and of rival behavior profoundly affects the empirical model by determining from which of several initial equilibria the impact of a policy change is measured. For example, one can use the standard symmetric duopoly market model to show five different equilibria resulting from different postulated behaviors of the duopolists (see Shubik and Levitan). Estimates of the impacts of a policy change such as trade liberalization on the duopoly market are conditional upon the position of the initial equilibrium. Thus, for a specific industry the researcher must know which behavior is correct if the empirical estimates are to accurately capture the impacts.

The empirical equilibrium is also dependent on the choice of payoff function. Alternative payoff functions considered in the oligopoly literature include maximization of market share, beat-the-average, and satisfying. Games using these functions will also have different solutions depending on conjectures of rival behavior. The builder of a realistic model of international grain traders would have several difficult choices and little data to guide such decisions.

Choice of Strategic Variable

In competitive and monopoly models little thought is devoted to selecting the strategic variable. Most of the empirical work presented by Richardson appears to use quantity. Choice of a strategic variable—price or quantity—by the researcher is another decision which alters the empirical results.

Consider a symmetric, non-cooperative game with product differentiation. Shubik and Levitan show that, in general, if price is the strategic variable, then the equilibrium price will be less than that which occurs if quantity is the strategic variable. This also means that given the same demand, the equilibrium quantity with price as the strategic variable will exceed that obtained when quantity is the choice variable.

Again the researcher's choice of strategic variable gives different initial equilibria from which the gains from trade liberalization are measured. As with the choice of payoff functions, little is known about the decision process for the large private international agricultural middlemen. The most popular guess about these firms is that quantity rather than price is their strategic variable.

Capacity

The ability of firms to supply the market at marginal cost (capacity) is important to oligopoly models. One can show that the existence of a binding capacity constraint by one duopolist gives more market power to the other duopolist.

In the context of international grain traders, whether firms can supply buyers in world markets at marginal cost is important to the empirical model. That is, do the few dominant firms in the industry have binding capacity constraints? If firms face capacity constraints market power may vary—being enhanced when capacity constraints of competing firms become binding.

Asymmetry

The models presented by Richardson also appear to assume symmetric firms. Symmetry among firms in an oligopoly model greatly facilitates the theoretical and empirical analysis. But casual observation of international grain traders suggests several sources of asymmetry. First, overhead costs and capacity for many items, such as port and terminal facilities and transportation, differ. Second, the marginal costs may not be symmetric. Third, a common observation is that largest grain traders have access to a superior network of information, market research, and contacts (see Morgan, 1979, for a story of how this information network operates). Finally, financing and management differ, especially when farmer cooperatives are trying to enter the export business (Reynolds 1980).

The introduction of asymmetry greatly complicates the analysis. How the equilibrium is affected depends upon where the asymmetry occurs and the structure of the model. For example, under profit maximization it is critical whether the asymmetry is introduced through fixed (overhead) costs or via marginal costs. Again the researcher needs to know much about the industry.

Interaction with the Government

For agricultural commodities many government or quasi-government boards have market power and are prone to use that power. Richardson notes that in contrast to the economics literature, much of the research in agricultural economics on imperfect competition has focused on the use of market power by these agents. A major omission of this literature is consideration of the interaction among governments with market power and large trading firms. Policy issues, such as land set-asides, export subsidies, and board sales handled by large traders, clearly raise the issue of interaction.

Some preliminary work in this area includes Bieri and Schmitz; Schmitz, McCalla, Mitchell, and Carter; and McCalla and Josling. This research is useful in framing some tentative hypotheses about firm-government interaction, but has not led to models of that process.

Implications

The implications of these observations for empirical modeling of imperfect competition in agricultural markets are sobering. Because of the sensitivity of the equilibrium to the structure of the model, the researcher must make critical decisions when formulating the model. Compared to modeling a market competitively, use of imperfect competition places a premium on a detailed knowledge of the industry. The researcher needs to invest sufficient resources to determine the payoff functions, the strategic variable, conjectures of rival behavior, the extent and nature of asymmetry, and the specification of marginal cost. The largest international agricultural trading firms are privately held and guard their secrets carefully. This means the basic data necessary to model these firms are not presently available. When the share of U. S. agricultural export controlled by the five largest firms is uncertain, the prospect of obtaining data sufficient to model imperfect markets is not promising.

References

Becker, G.S. 1965. A Theory of the Allocation of Time. *The Economic Journal.* 75:493-517.

Bieri, J. and A. Schmitz. 1974. Market Intermediaries and Price Instability: Some Welfare Implications. *American Journal of Agricultural Economics.* 56:280-285.

Lancaster, K. 1966. A New Approach to Consumer Theory. *Journal of Political Economy.* 74:132-157.

McCalla, A. F. and T.E. Josling. 1981. *Imperfect Markets In Agricultural Trade.* Montclair, NJ: Allanheld, Osum.

Morgan, D. 1979. *Merchants of Grain.* New York: Penguin Books.

Reynolds, B. 1980. *Producers Export Company: The Beginnings of Cooperative Grain Exporting.* Farmer Cooperative Research Report No. 15. Economics Statistics, and Cooperatives Service, U. S. Department of Agriculture. Jan.

Schmitz, A., A.F. McCalla, D.O. Mitchell, and C. Carter. 1980. *Grain Export Cartels.* Cambridge, MA: Ballinger Publishing Company.

Shubik, M. with R. Levitan. 1980. *Market Structure and Behavior.* Cambridge, MA: Harvard Univ. Press.

4

Strategic Trade Theory and Agricultural Markets: An Application to Canadian and U.S. Wheat Exports to Japan

Marie C. Thursby and Jerry G. Thursby

Introduction

International trade theory has changed dramatically in the past decade. Its focus has been on imperfect competition, and its policy prescriptions have called into question the traditional free trade "ethic." With perfect competition, trade policy can be first-best policy only for a large country to improve its terms of trade, and even then, free trade is preferable from a world perspective. With imperfect competition, firms do not take price as given, and they realize their actions, as well as those of their rivals, will affect market outcomes. This leads to all sorts of possibilities for strategic behavior by firms. It also gives rise to the possibility for governments to use policy to alter this rivalry to the benefit of their firms or consumers. This has been the focus of the literature on strategic trade policy.

There has been much interest in this work because of the potential for governments to alter the strategic positions of their firms in world markets. At the same time, academic economists have been quick to caution that much more must be known about market parameters and the nature of imperfect competition before results are practically applicable. This is not an idle caution because policy prescriptions vary from subsidies to taxes depending on such factors as demand parameters, the number of firms competing, and the way in which they compete.

In this paper we examine the relevance of this type of analysis for agricultural markets. It is easy to argue that agricultural production is perfectly competitive, but large marketing institutions are prevalent in these markets. Government marketing boards are found on both the export and import side of agricultural markets. For example, several major exporters of dairy products and grain sell through marketing boards, and major importers of grains, tobacco, and silk purchase through them (Hoos 1979, Kostecki 1982).[1] It is also not uncommon to find highly concentrated private export industries. Nutmeg, orange juice (FCOJ), and grains provide a few examples. Rodrik (1989) reports that one firm accounts for 20-45 percent of Indonesian nutmeg exports, with Indonesia supplying three-fourths of the world market. It is estimated (ITC 1984) that three firms account for 85 percent of Brazilian FCOJ exports, with Brazil accounting for 90 percent of U.S. FCOJ imports. Since the large U.S.S.R. purchases of grain in the mid-1970s, the competitiveness of the U.S. grain exporting industry has been highly disputed. Estimates for 1974/75 (Conklin 1982) are that the largest four exporters accounted for 61 percent of U.S. wheat exports.

We present an agricultural trade model which incorporates these types of institutions. Our aim is to show how this type of model can be used to infer the nature of competition in the export market as well as for policy analysis.

In our model, two countries export a (perfectly) competitively produced product. One of the countries exports its product through a marketing board while in the other, the export industry is composed of large private firms. Rather than specify the nature of rivalry among these firms and the board, we use conjectural variation parameters to allow for a range of competitive assumptions (including Cournot and Bertrand behavior). Given information on demand parameters, the model can be calibrated to market data to determine values for these conjectures.

Although models of this type could apply to any commodity, our assumptions about marketing institutions are meant to comply with stylized facts from the world wheat market. Therefore we shall call the product wheat, and in Section III we report results from calibrating the model to data for Canadian and U.S. exports of wheat to Japan. The wheat market is a natural one to consider given the long history of oligopoly models for world wheat trade (for examples, see McCalla 1966; Taplin 1969; Alaouze et al. 1978; Schmitz et al. 1981; Karp and McCalla 1983; Paarlberg and Abbott 1984; Kolstad and Burris 1986; and

Thursby 1988). With the exception of Thursby (1988), these studies tend to abstract from firm behavior and focus on countries as agents with market power. Studies have made a variety of assumptions about numbers of countries and the nature of competition (price or quantity), but it is worth noting that McCalla's original study found price data to be consistent with a Canadian-United States duopoly. Kolstad and Burris' spatial equilibrium analysis in which producing country governments are Nash quantity competitors finds that a U.S.-Canadian duopoly comes the closest to predicting actual trade for 1972/73.

Our calibration exercise is an example of the type of analysis done by Dixit (1987), Rodrik (1988), Venables and Smith (1986), Baldwin and Krugman (1987), Krishna et al. (1989), among others. Helpman and Krugman (1989) and Richardson (1989) provide useful surveys of this type of work.

Finally, we note the methodological limitations of this type of analysis. Conjectural variations is a short-cut method of parameterizing the nature of competition, where Cournot, Bertrand, and competitive behavior are special cases. The calibration of such a model to market data does not represent a true test of theory, but merely gives an indication of parameters consistent with market data. In the current state of modelling technology, this remains the most tractable way to make inferences about market behavior with imperfect competition. As we note in Section III, our study differs from the bulk of calibration exercises by our attempt to determine variances of our estimates.

The Model

Consider a world in which two countries export wheat to a third country. Each of the exporting countries consumes the good, but because of restrictions outside the model, they do not import it. This could be explained by quotas, but for simplicity we abstract from them here. The good (wheat) is competitively produced, and producers sell to a distributor or marketing agent rather than directly to consumers. In practice this might occur because of technological features of transportation and marketing services, but, again, we abstract from these here. For simplicity we also abstract from inventory decisions. The competitive producer supply curve is upward sloping.

We call the two exporting countries C and U. In country C a statutory marketing board is the sole marketing agent. This board handles all domestic, as well as foreign, sales to consumers. In country U, distributors are private firms. There are m such firms, each assumed to maximize profits. Country C's marketing board, on the other hand, is assumed to maximize the joint returns of its competitive producers plus export revenue.[2]

The timing of actions is as follows. The marketing board and exporting firms maximize their objective functions, given their assumptions, i.e., conjectures, about each others behavior, and take as given the competitive supply curves, any taxes or subsidies of their respective governments, and any import tariffs levied by the government in the third market. Hence, governments are assumed to be able to precommit to their policies. Throughout the paper the analysis will be partial equilibrium.

Demand

All wheat produced in country C is homogeneous, as is all wheat in country U. However, the wheat of the two countries need not be perfect substitutes. We denote the import country by J. If we were focusing on the world market as a Canadian-United States duopoly, demand in J would refer to world demand for exports of C and U. Alternatively, we could focus on a single country's market.

Let P^{ri} refer to the consumer price in country J of wheat from country i. In the absence of a tariff, this equals the cif price of country i's wheat, P^{ji}. With an ad valorem tariff, $P^{ri} = (1+t)P^{ji}$, while a specific tariff gives $P^{ri} = P^{ji} + t$, where t denotes the tariff. Let X^i denote the quantity of wheat exported by country i to j. The demand functions in J are assumed to be

$$X^c = A^c - B^c P^{rc} + K P^{ru} \tag{1}$$

$$X^u = A^u + K P^{rc} - B^u P^{ru} \tag{2}$$

where all parameters are positive and $B^c B^u - K^2 > 0$. In inverse form these functions are given by

$$P^{rc} = a^c - b^c X^c - k X^u \tag{3}$$

$$P^{ru} = a^u - k X^c - b^u X^u \tag{4}$$

where all parameters are positive and $b^c b^u - k^2 > 0$. This system has the virtue of being consistent with estimated linear functions. It can be derived from quadratic utility[3]

$$U(X^c, X^u, E) = a^c X^c + a^u X^u - .5 \left[b^c (X^c)^2 + b^u (X^u)^2 + 2k X^c X^u \right] + E \tag{5}$$

where E is expenditure on all other goods.

Since we assume the exporting countries consume only their own wheat, the comparable utility would be

$$U(Y^i) = d^i Y^i - .5e^i(Y^i)^2 \qquad (6)$$

where Y^i denotes domestic consumption for $i = U, C$ and parameters are positive. Domestic inverse demands in the exporting countries are then

$$P^{ii} = d^i - e^i Y^i. \qquad (7)$$

Marketing Board's Problem

The marketing board maximizes the joint returns of competitive producers in its country plus export revenue. For ease of exposition we shall interpret J as the world market, so that the board's sales are either for domestic or export use. If J were to denote a single country, then total board returns would include the sum of revenues from all export markets.

In our case the board's objective function is given by

$$\left[P^{cc}(Y^c) + cs^c \right] Y^c + \left[P^{jc}(X^c, X^u) + es^c - c^c \right] X^c - F$$

$$- \int_0^{Y^c + X^c} \left[s^c(q) - ps^c \right] dq \qquad (8)$$

where inverse demands are defined above, $s^c(Y^c + X^c)$ is the competitive producer supply price, cs^c is a consumer subsidy, es^c is an export subsidy, ps^c is a producer subsidy, c^c is transport cost for export sales, and F is fixed cost. We shall presume that operating and domestic marketing costs are included in F.

In the absence of any regulations, the board's domestic and export sales would be determined by the following first order conditions:

$$P^{cc}(1 + e_c^{cc}) = s^c - (ps^c + cs^c) \qquad (9)$$

and

$$P^{jc}(1 + e_c^{jc} + e_u^{jc} \, v^{cu} X^c / X^u) = s^c + c^c - (ps^c + es^c) \qquad (10)$$

where e_c^{cc} is the domestic inverse demand elasticity (price flexibility), e_c^{jc} is the own inverse import demand elasticity in country j, e_u^{jc} is the cross elasticity of inverse demand for imports from C with respect to X^u, and v^{cu} is the board's conjecture about the response of X^u to a change in the board's exports. Throughout the paper elasticities will carry their natural sign. Under the assumption of Cournot competition, the board considers U.S. exports as given, or $v^{cu} = 0$. With Bertrand competition, the board would take the U.S. export price as given. Given our demand structure this means the conjecture about dX^u/dX^c is $-k/b^u$ with Bertrand.

Many marketing boards are regulated in their domestic pricing. Just, et al. (1979), Markusen (1984), Thursby (1988), and Krishna and Thursby (1988) have pointed out in several contexts that such regulation will affect export decisions. Hence we consider an alternative specification of the board's problem.

With a regulation that the board equate domestic demand and supply prices, it's maximization problem would be constrained by

$$P^{cc}(Y^c) + cs^c = s^c(Y^c+X^c) - ps^c. \tag{11}$$

Given this constraint and the board's conjecture about the effect of its sales on exports of firms in country U, the first order conditions for the marketing board are (11) and

$$P^{cc}e_c^{cc} + P^{jc}\phi\left[1+e\,\xi^c+ e_u^{jc}v^{cu}X^c/X^u\,\right] = \phi(s^c+c^c-es^c-ps^c) \tag{12}$$

where $\phi = \left[(\partial P^{cc}/\partial Y)/\partial s^c/\partial(X^c+Y^c)\right]-1$.

Notice from (9) and (10) that without the domestic price regulation, the board equates perceived marginal revenue in each market with the competitive supply price net of taxes and subsidies. The term "perceived" is used because of the conjectural variation parameter. From (11) and (12) it is apparent that the regulation prevents the board from equating these two.

Export Industry of Country U

Modelling country U's export industry poses more problems than does the marketing board. In a conjectural variations framework, each firm in U's industry will have a conjecture about the board's behavior, a conjecture about other firms' behavior in the export market, and a conjecture about rival firms' behavior at home. *A priori*, there is no reason for these conjectures to be the same. One way to avoid this type of problem is to assume no domestic sales by exporters, but that is clearly unrealistic in the case of U.S. wheat, for example. Another way to simplify the problem is to adopt a model similar to that of Thursby (1988) in which there are two types of marketing firms, one which exports and one which sells only in the domestic market because of a cost disadvantage. In the limit the model allows the possibility of imperfect competition in the export sector, but these firms cannot exercise oligopoly power in the domestic market if there is a competitive fringe of firms who market the good domestically.

This approach is consistent with evidence from the U.S. market. Conklin (1982) reports lower concentration ratios for domestic grain sales than for export sales. Caves and Pugel (1982) present similar evidence based on a survey of members of the North American Export Grain Association. Their evidence points to the largest firms handling a majority of "direct" export sales, while many smaller firms purchase grain from farmers to sell domestically or to the largest exporters who then export it (the latter type of sale being classified as "indirect" exports).

Suppose there are m = n+h firms, the last h of which have a cost disadvantage relative to the first n firms. Profit for the ith firm is given by

$$\pi_i = P^{uu}(Y^u)y_i^u + P^{ju}(X^c,X^u)x_i^u - s^u(Y^u + X^u)(x_i^u+y_i^u)$$
$$- F_{ix} - F_y + (ps^u+cs^u)y_i^u + (ps^u+es^u-c^u)x_i^u \qquad (13)$$

where $Y^u = \sum_{i=1}^{m} y_i^u$, $X^u = \sum_{i=1}^{m} x_i^u$, y_i^u is domestic sales of firm i, x_i^u is export sales of firm i, F_{ix} is fixed cost associated with export activity, F_y is fixed cost associated with domestic operation, ps^u is a producer subsidy, cs^u is a consumer subsidy, es^u is an export subsidy, and c^u is per unit transport cost to export to J. Firms are differentiated only by the export fixed cost parameter, F_{ix}, and for simplicity we assume it takes on only two values, low (F_{1x}) or high (F_{2x}). For i = 1,...,n, $F_{ix} = F_{1x}$ and for i = n+1,..., m, $F_{ix} = F_{2x}$.[4]

We assume free entry and a low enough value for F_y that there are many firms in the domestic market. Thus we assume competitive conjectures in the domestic market, so that firm's first order conditions for domestic sales give P^{uu} equal to the competitive supply price (net of producer and consumer subsidies). However, for high enough values of F_{2x}, only type 1 firms will enter the export market. Under our assumptions all type one firms are identical. We look at a symmetric equilibrium in U's exports, in which case a representative firm's first order condition for exports can be written as

$$P^{ju}\left[\; 1 + (e_{u}^{j\,u}/n)(1 + v^{uu}) + e_{c}^{j\,u}v^{uc}x_{i}^{u}/X^{c}\;\right] =$$
$$s^{u} + c^{u} - ps^{u} - es^{u} + \Psi x_{i}^{u}(1 + v^{uu}) \tag{14}$$

where $\Psi = \partial s^{u}/\partial(Y^{u} + X^{u})$, $e_{u}^{j\,u}$ is country j's inverse demand elasticity with respect to X^{u}, $e_{c}^{j\,u}$ is the cross elasticity of inverse demand for U's wheat with respect to X^{c}, v^{uu} is the representative firm's conjecture about responses of all other type 1 firms in the export market, v^{uc} is the representative firms's conjecture about the response of C's board in the export market. Hence each firm equates perceived marginal revenue from exports with perceived marginal cost. Notice that Ψ allows for the possibility of oligopsony power by exporters.

Equilibrium and Policy Analysis

The Nash equilibrium to this game among country C's board and country U's firms is determined by simultaneous solution of the demand system and the relevant first order conditions. The board's relevant first order conditions are (9) and (10) if it is unregulated, and they are given by (11) and (12) if it is regulated. There are n equations given by (14) for country U plus an equation for U's domestic demand equal supply. These equations plus j's import demand functions give the equilibrium levels of domestic sales and exports of each country.

Recall that the relevant first order conditions assume given levels of t, ps^{i}, es^{i}, and cs^{i}. Comparative statics exercises can be done to show how equilibrium exports and prices vary with changes in these parameters. In addition, if we specify the government objectives in the importing and exporting countries, we can solve for optimal levels of these policies. In this model, the importing country has market power. This gives the potential for its government to levy a non-zero optimum tariff/subsidy. The nature of the optimal import policy will depend on the government's objective function as well as the nature of competition among exporters. Optimal policies for the exporting countries' governments will vary depending on the nature of competition and the presence or absence of regulation. The number of private firms competing, as well as whether price or quantity is the strategic variable, will be important in determining optimal policy.[5] These factors are also important in assessing the impact of trade liberalization.

Calibration of the Model

As an illustration of the use of this type of model to indicate market structure, we now turn to a calibration exercise. Data are from the market for Japanese wheat imports from the United States and Canada. Together the United States and Canada account for around 80 percent of Japan's wheat imports. Canada exports through the Canadian Wheat Board and the U.S. export industry is largely composed of private firms. Data for the mid-1970s to the present show that, depending on the year, the number of U.S. firms exporting ranges from around 30 to around 60 with the largest four U.S. firms accounting for 60 percent of U.S. wheat exports in 1974/75.

Since there is no information on market structure from the domestic first order conditions we shall focus on the first order conditions for exports for both U.S. firms and the Canadian board. Since exports to Japan are a small portion of total wheat production in both countries we shall assume that both the board and U.S. exporters ignore any potential effects of exports on the domestic supply price. That is, for the purposes of this exercise, ϕ and Ψ are assumed to be zero. Our purpose is to use available data to estimate values for v^{uu}, v^{uc} and v^{cu}.

The relevant Canadian equation is (10) which we rewrite for convenience as

$$v^{cu} = - X^u \left[\mu^c + P^{jc} e_{\ell}^{jc} \right] / e_{\ell}^{jc} P^{jc} X^c \tag{15}$$

where

$$\mu^c = P^{jc} - s^c - c^c + es^c + ps^c$$

and X^c and X^u now denote exports of Canada and the United States to Japan. For any given year, μ^c, P^{jc}, X^u and X^c are observable. Unobserved parameters are e_{ℓ}^{jc} and $e_{\ell\ell}^{jc}$; however, estimates for these can be obtained and we can solve for v^{cu}.

For the United States there are n first-order conditions for exports given by (14) with $\Psi = 0$ (since U.S. exporters are assumed to ignore domestic supply price effects of their actions). Since there are two conjectures in (14) it will be necessary to condition on one and estimate the other. We begin by conditioning on v^{uu}. Summing over all n firms we can write v^{uc} as

$$v^{uc} = -X^c \left[n\mu^u + P^{ju} e_{\ell}^{ju}(1+v^{uu}) \right] / e_{\ell}^{ju} P^{ju} X^u \tag{16}$$

where

$$\mu^u = P^{ju} - s^u - c^u + ps^u + es^u.$$

For any given year, n, μ^u, P^{ju}, X^u and X^c are observable and we condition on $v^{uu} \cdot e_u^{ju}$ and e_{ζ}^{ju} are unobserved but estimable parameters, hence v^{uc} can be estimated.

Conversely, we can condition on v^{cu} which gives

$$v^{uu} = -\left[n\mu^u + P^{ju}(e_u^{ju} + e_{\zeta}^{ju} v^{uc} X^u / X^c)\right]/P^{ju} e_u^{ju}. \qquad (17)$$

Conditioning on v^{cu} and using estimates of e_u^{ju} and e_{ζ}^{ju} we can estimate v^{uu}.

Estimates of the parameters e_{ζ}^{jc}, e_u^{jc}, e_u^{ju} and e_{ζ}^{ju} can be obtained in a variety of ways.[6] We estimate demand in its direct rather than inverse form. To be consistent with equations (1) and (2) our estimated equations are linear functions of prices. Based on the estimated parameters of the direct demand equations we calculate the inverse elasticities using the relationship between the parameters of (1) and (2) with (3) and (4). These estimates are then used in equations (15), (16) and (17). In addition, the estimated variances of the elasticities are used in approximating the variances of the estimated conjectures. Details of our procedure are in Appendix A. Data sources are given in Appendix B.

Finally, notice that the number of U.S. firms enters the equations for v^{uu} and v^{uc}. Depending on the year, between 30 to 60 U.S. firms export wheat. However, using these numbers for n would be inappropriate, since our equations implicitly assume the export industry is composed of symmetric firms. The industry is clearly not symmetric, and ideally one would adopt a model which endogenized the size distribution of firms. Since that is not a tractable problem, we follow current best practice and compute the Herfindahl equivalent number of symmetric firms. That is, $n = 1/H$ where $H = \sum_{i=1}^{N} s_i^2$, N is the actual number of firms, and s_i is the share in exports of the ith firm.[7] For the years we consider, the Herfindahl ranged between .07 and .11, which implies equivalent numbers of firms between 9 and 14. Our results are not sensitive to changes in this range, and we report results based on 10 firms.

Tables 4-1 to 4-4 contain results for the years 1977/78, 1979/80, 1981/82 and 1983/84 as well as the own and cross-price elasticities. Notice in all years Japan's import demands are quite inelastic. This may well reflect the fact that wheat is purchased through the Japanese Food Agency. The implied conjectures are remarkably constant across years. The Canadian conjectures about U.S. firms vary from -.908 to -.943. These are very close to Bertrand conjectures as, given our elasticities, Bertrand is around -.85. Cournot conjectures would be zero. The estimated conjecture for 1977/78 is significantly different from zero at a 6 percent significance level. The conjectures for 1979/80, 1981/82 and 1983/84 are significantly different from zero at the 14 percent, 12 percent and 12 percent levels, respectively.

Interpretation of results for v^{uu} and v^{uc} is a bit trickier. As should be the case, point estimates are the same whether we condition on v^{uu} or v^{uc}; however, variances are different. If U.S. firms have Cournot conjectures about other U.S. firms (that is, $v^{uu} = 0$), the implied conjectures about the Canadian Board are closer to Bertrand conjectures. In 1977/78 the implied v^{uc} conditional on $v^{uu} = 0$ is -1.219 and the Bertrand conjecture for that year is -0.994. For 1979/80, 1981/82 and 1983/84 the implied v^{uc} and Bertrand conjectures are (-1.14, -1.052), (-1.169, -1.047) and (-1.159, -1.069), respectively. Unfortunately, the standard errors are somewhat large. More competitive conjectures of U.S. firms about each other are -.5 and -.9 and in both cases implied conjectures about the board are less competitive than Bertrand. Note that the standard errors decline as v^{uu} declines. Finally, we do not condition on $v^{uu} = -1$ since that is the limiting case of perfect competition.[8] In that case excess supply would describe U.S. exports and v^{uc} would be irrelevant.

Table 4-1. Conjectural Results for 1977/1978[a]

Elasticities[b]			
ε_u^{ju}	ε_c^{jc}	ε_c^{ju}	ε_u^{jc}
-0.099 (.144)	-0.340 (.301)	0.110 (.150)	0.255 (.347)

Canadian Results	
v^{cu}	
-0.943	(0.484)

U.S. Results				
v^{uc} Conditional on v^{uu}			v^{uu} Conditional on v^{uc}	
v^{uu}	v^{uc}		v^{uc}	v^{uu}	
-0.900	-0.099	(0.123)	-0.994	-0.229	(1.224)
-0.500	-0.632	(1.027)	-0.500	-0.599	(0.567)
0.000	-1.299	(2.173)	0.000	-0.974	(0.109)

[a]Standard errors in parentheses

[b] $\varepsilon_u^{ju} = (\partial X^u/\partial P^u)(P^u/X^u)$ $\varepsilon_c^{jc} = (\partial X^c/\partial P^c)(P^c/X^c)$

$\varepsilon_c^{ju} = (\partial X^u/\partial P^c)(P^c/X^u)$ $\varepsilon_u^{jc} = (\partial X^c/\partial P^u)(P^u/X^c)$

Table 4-2. Conjectural Results for 1979/1980[a]

Elasticities[b]

ε_u^{ju}	ε_c^{jc}	ε_c^{ju}	ε_u^{jc}
-0.149 (.218)	-0.506 (.448)	0.177 (.240)	0.359 (.489)

Canadian Results

v^{cu}

-0.921   (0.624)

U.S. Results

v^{uc} Conditional on v^{uu}			v^{uu} Conditional on v^{uc}		
v^{uu}	v^{uc}		v^{uc}	v^{uu}	
-0.900	-0.101	(0.090)	-1.052	-0.077	(1.404)
-0.500	-0.563	(0.801)	-0.500	-0.554	(0.558)
0.000	-1.140	(1.780)	0.000	-0.987	(0.220)

[a]Standard errors in parentheses

[b] $\varepsilon_u^{ju} = (\partial X^u/\partial P^u)(P^u/X^u)$ $\qquad$ $\varepsilon_c^{jc} = (\partial X^c/\partial P^c)(P^c/X^c)$

$\varepsilon_c^{ju} = (\partial X^u/\partial P^c)(P^c/X^u)$ $\qquad$ $\varepsilon_u^{jc} = (\partial X^c/\partial P^u)(P^u/X^c)$

Conclusion

In this paper we have presented an imperfectly competitive trade model which incorporates the types of institutions prevalent in agricultural marketing. A virtue of this model is its flexibility in describing a variety of market structures, to include highly competitive ones. Given estimated demand parameters and market data, it can be used to make inferences about the nature of competition. The model and "calibrated" estimates of behavior can then be used for policy experiments. In this paper we have concentrated on the calibration exercise, but our ongoing work will focus on trade liberalization and other policy experiments.

Our work differs in several respects from other calibration studies. By incorporating both private firms and a marketing board we have allowed more asymmetries on the exporting side than have other studies. Our board and firms have different objective functions. In addition we allowed conjectures to be asymmetric rather than focusing on aggregate conjectures.[9] Finally, our method of examining sensitivity differs from other calibration studies. We used the estimated variances and covariances of demand elasticities to approximate the vari-

Table 4-3. Conjectural Results for 1981/1982[a]

Elasticities[b]			
ε_u^{ju}	ε_c^{jc}	ε_c^{ju}	ε_u^{jc}
-0.128 (.187)	-0.458 (.405)	0.153 (.209)	0.320 (.435)

Canadian Results

v^{cu}	
-0.918	(0.584)

U.S. Results

v^{uc} Conditional on v^{uu}			v^{uu} Conditional on v^{uc}		
v^{uu}	v^{uc}		v^{uc}	v^{uu}	
-0.900	-0.097	(0.093)	-1.047	-0.103	(1.353)
-0.500	-0.574	(0.821)	-0.500	-0.562	(0.540)
0.000	-1.169	(1.828)	0.000	-0.982	(0.216)

[a]Standard errors in parentheses

[b] $\varepsilon_u^{ju} = (\partial X^u/\partial P^u)(P^u/X^u)$ $\qquad \varepsilon_c^{jc} = (\partial X^c/\partial P^c)(P^c/X^c)$

$\varepsilon_c^{ju} = (\partial X^u/\partial P^c)(P^c/X^u)$ $\qquad \varepsilon_u^{jc} = (\partial X^c/\partial P^u)(P^u/X^c)$

ances of estimated conjectures. The usual practice in calibration studies is to examine the effects of arbitrary changes in parameters.

Our work differs from other studies of imperfect competition in international agricultural trade because most other studies focus on the market power of governments rather than firms. Our results suggest that the Canadian-United States rivalry in the Japanese wheat market is more competitive than Cournot competition. To see this recall that Cournot competition would be the result if $v^{uu} = v^{uc} = v^{cu} = 0$. Canadian conjectures were not zero in any years. Because of our having to estimate one of the U.S. conjectures conditional on the other, we have a range of estimates for the U.S. conjectures. But notice, that none of the combinations give $v^{uu} = v^{uc} = 0$.

Finally, we should note that our results are driven by the inelastic demand for wheat by the Japanese Food Agency. In future work, we intend to examine alternative ways of modelling Japanese demand and the sensitivity of the analysis to different specifications.

Table 4-4. Conjectural Results for 1983/1984[a]

Elasticities[b]			
ε_u^{ju}	ε_u^{jc}	ε_c^{ju}	ε_u^{jc}
-0.116 (.170)	-0.421 (.373)	0.142 (.194)	0.289 (.393)

Canadian Results

v^{cu}

-0.908   (0.570)

U.S. Results

v^{uc} Conditional on v^{uu}			v^{uu} Conditional on v^{uc}		
v^{uu}	v^{uc}		v^{uc}	v^{uu}	
-0.900	-0.098	(0.116)	-1.069	-0.076	(1.350)
-0.500	-0.570	(0.763)	-0.500	-0.559	(0.496)
0.000	-1.159	(1.750)	0.000	-0.983	(0.271)

[a]Standard errors in parentheses

[b] $\varepsilon_u^{ju} = (\partial X^u/\partial P^u)(P^u/X^u)$ $\qquad$ $\varepsilon_c^{jc} = (\partial X^c/\partial P^c)(P^c/X^c)$

$\varepsilon_c^{ju} = (\partial X^u/\partial P^c)(P^c/X^u)$ $\qquad$ $\varepsilon_u^{jc} = (\partial X^c/\partial P^u)(P^u/X^c)$

Notes

1. For OECD trade in 34 agricultural products for 1976, Kostecki (1982, pp. 26, 286-8) estimates 28 percent of exports and 27 percent of imports are accounted for by state trading.

2. See Markusen (1984) for an analysis of marketing boards which maximize profits and Veeman and Lyons (1982) and Veeman (1987) for a discussion of other market board objectives.

3. See Thursby and Thursby 1988 on issues related to whether demand for agricultural commodities is assumed to come from utility or profit maximization. It is also worth noting that this demand system is the same form as that of Dixit (1987) and Laussel *et al.* (1988).

4. Thursby (1988) differed by assuming unit variable costs to vary among firms. Allowing different fixed costs has the same impact and is closer in spirit to the types of differentiation reported by Caves and Pugel (1982).

5. For an analysis of the number of firms and effects of regulation on optimal policies see Thursby (1988). See Krishna and Thursby (1988)

for a general analysis of these effects as well as issues of market segmentation and the strategic variable. Thursby differs from the work here by assuming Cournot competition and linear demand and competitive supply curves. This allows for explicit solutions, so that she examines Nash equilibria in which both exporting country governments are active in policy.

6. See Dixit (1987) for an alternative procedure.

7. If the industry were symmetric, the Herfindahl would be $\sum_1^N (1/N)^2$, so that $1/H$ would be the number of firms in the industry.

8. This is because wheat sold by different U.S. firms is modeled as homogenous. With a homogenous good, the only symmetric equilibrium with Bertrand competition is the competitive one. Incorporating inventories or capacity choices would allow positive profits with Bertrand competition. Similarly, an asymmetric industry with Bertrand competition need not lead to the limiting case of perfect competition.

9. Some studies focus on aggregate conjectures (e.g., Dixit 1987 and Laussel et. al. 1988), while others do not (see e.g., Baldwin and Krugman 1987 and Krishna et al. 1989).

References

Alaouze, C. M. and A.S. Watson, and N. H. Sturgess. 1978. Oligopoly Pricing in the World Wheat Market. *American Journal of Agricultural Economics.* 60:173-85.

Baldwin, R. and P. Krugman. 1987. Market Access and International Competition: A Simulation Study of 16K Random Access Memories. In *Empirical Methods for International Trade*, ed. Robert Feenstra. Cambridge: MIT Press.

Carter, C. and A. Schmitz. 1979. Import Tariffs and Price Formation in the World Wheat Market. *American Journal of Agricultural Economics.* 61:517-22.

Caves, R. E. 1978. Organization, Scale, and Performance in the Grain Trade. *Food Research Institute Studies.* 16:107-23.

Caves, R. E. and T.A. Pugel. 1982. New Evidence on Competition in the Grain Trade. *Food Research Institute Studies.* 18:261-74.

Conklin, N.C. 1982. An Economic Analysis of the Pricing Efficiency and Market Organization of the U.S. Grain export system. Staff Study GAO/CED-82-61S. Washington: U.S. General Accounting Office.

Dixit, A. K. 1984. International Trade Policy for Oligopolistic Industries. *Economic Journal* 94:1-16.

—— 1987. Optimal Trade and Industrial Policies for the U.S. Automobile Industry. In *Empirical Methods for International Trade*, ed. Robert Feenstra. Cambridge: MIT Press.

Eaton, J. and G.M. Grossman. 1986. Optimal Trade and Industrial Policy Under Oligopoly. *Quarterly Journal of Economics*. 101:383-406.

Gallagher, P. M., M.B. Lancaster, M. Bredahl, and T.J. Ryan. 1981. *The U.S. Wheat Economy in an International Setting: An Econometric Investigation*. Technical Bulletin no. 1644. Washington: Economics and Statistics Service, U.S. Department of Agriculture.

Helpman, E. and P.R. Krugman. 1989. *Trade Policy and Market Structure*. Cambridge: MIT Press.

Hjort, K. 1988. Class and Source Substitutability in the Demand for Imported Wheat. Ph.D. diss., Purdue University.

Hoos, S. 1979. *Agricultural Marketing Boards: An International Perspective*. Cambridge, Mass.: Ballinger.

Just, R., A. Schmitz, and D. Zilberman. 1979. Price Controls and Optimal Export Policies Under Alternative Market Structures. *American Economic Review*. 69:706-15.

Karp, L. S. and A.F. McCalla. 1983. Dynamic Games and International Trade: An Application to the World Corn Market. *American Journal of Agricultural Economics*. 65:641-56.

Kolstad, C.D. and A.E. Burris. 1986. Imperfectly Competitive Equilibria in International Commodity Markets. *American Journal of Agricultural Economics*. 68:25-36.

Kostecki, M.M. 1982. State Trading in Agricultural Products by the Advanced Countries. In *State Trading in International Markets*, ed. M. M. Kostecki. New York: St. Martin's Press.

Krishna, K., K. Hogan, and P. Swagel. 1989. The Non-optimality of Optimal Trade Policies: The U.S. Automobile Industry Revisited, 1979-1985. Mimeo.

Krishna, K. and M. Thursby. 1988. *Optimal Policies with Strategic Distortions*. NBER Working Paper no. 2527. Cambridge, Mass.: National Bureau of Economic Research.

Laussel, D., C. Montet and A. Peguin-Feissolle. 1988. Optimal Trade Policy Under Oligopoly. *European Economic Review*. 32:1547-65.

Markusen, J.R. 1984. The Welfare and Allocative Effects of Export Taxes versus Marketing Boards. *Journal of Development Economics* 14:19-36.

McCalla, A.F. 1966. A Duopoly Model of World Wheat Pricing. *Journal of Farm Economics*. 48:711-27.

McCalla, A.F. and T.E. Josling. 1981. *Imperfect Markets in Agricultural Trade*. Montclair, NJ: Allanheld, Osmun.

McCalla, A.F. and A. Schmitz. 1982. State Trading in Grain. In *State Trading in International Markets*, ed. M. M. Kostecki. New York: St. Martin's Press.

Paarlberg, P. L. and P.C. Abbott. 1984. Towards a Countervailing Power Theory of World Wheat Trade. In *International Agricultural Trade*, eds. G.G. Storey, A. Schmitz, and A.H. Sarris. London: Westview Press.

Richardson, J.D. 1988. *Empirical Research on Trade Liberalization with Imperfect Competition: A Survey.* NBER Working Paper no. 2883. Cambridge, Mass: National Bureau of Economic Research.

Rodrik, D. 1988. Imperfect Competition, Scale Economies and Trade Policies in Developing Countries. In *Trade Policy and Empirical Analysis*, ed. R. E. Baldwin. Chicago: University of Chicago Press.

Schmitz, A. and A.F. McCalla. 1979. The Canadian Wheat Board. In *Agricultural Marketing Boards*, ed. S. Hoos. Cambridge, Mass.: Ballinger.

Schmitz, A., A.F. McCalla, D.O. Mitchell, and C. Carter. 1981. *Grain Export Cartels.* Cambridge, Mass: Ballinger.

Thompson, R. L. 1978. *Japanese Import and Pricing Policies.* U.S. Department of Agriculture.

Thursby, M. 1988. Strategic Models, Market Structure and State Trading: An Application to Agriculture. In *Trade policy and Empirical Analysis* R. E. Baldwin, ed. Chicago: University of Chicago Press.

Thursby, J.G. and M.C. Thursby. 1988. Elasticities in International Trade: Theoretical and Methodological Issues. In *Elasticities in International Agricultural Trade* C. A. Carter and W. H. Gardiner, eds. Boulder: Westview Press.

U.S. General Accounting Office. 1982. *Market Structure and Pricing Efficiency of U.S. Grain Export System.* GAO/CED-82-61. Washington.

U.S. International Trade Commission. 1984. *Frozen Concentrated Orange Juice from Brazil.* U.S. International Trade Commission Publication no. 1623. Washington.

Veeman, M. and A. Loyns. 1988. Agricultural Marketing Boards in Canada. In *State Trading in International Markets.* M.M. Kostecki, ed. New York: St. Martin's Press.

Veeman, M. 1987. Marketing Boards: The Canadian Experience. *American Journal of Agricultural Economics.* 992-1000.

Venables, A. and A. Smith. 1986. Trade and Industrial Policy Under Imperfect Competition. *Economic Policy.* 1:622-672.

APPENDIX A

Procedure for Estimating v^{cu}, v^{uc} and v^{uu}

Referring to equations (15), (16) and (17), the unknown parameters are e_ζ^{jc}, e_u^{jc}, e_u^{ju} and e_ζ^{ju}. As noted we obtain estimates by first estimating the direct demand system given by equations (1) and (2). Since (1) and (2) implicitly hold constant other determinants of demand, we proceed as follows. We estimated the two equation linear regression systems

$$X_t^u = \beta_0^u + \beta_1^u P_t^{ju} + \beta_2^u P_t^{jc} + \beta_3^u Jp_t^u + \beta_4^u Inc_t + \beta_5^u Stks_t + \beta_6^u Strike_t + \varepsilon_t^u \qquad (A1)$$

$$X_t^c = \beta_0^c + \beta_1^c P_t^{ju} + \beta_2^c P_t^{jc} + \beta_3^c Jp_t^c + \beta_4^c Inc_t + \beta_5^c Stks_t + \varepsilon_t^c \qquad (A2)$$

X	=	the level of Japanese per capita imports,
P^{ju}	=	the real import price of U.S. wheat in yen,
P^{jc}	=	the real import price of Canadian wheat in yen,
Jp	=	the Japanese resale price set by the Japanese Food Agency,
Inc	=	per capita real Japanese income,
Stks	=	Japanese per capita beginning stocks + production – exports,
Strike	=	variable to reflect U.S. west coast dock strike activity,

and the superscripts refer to Canadian or U.S. data and coefficients. Data are annual from 1960 to 1985. To account for possible correlation of disturbances between the two regression regimes we used the seemingly unrelated regression procedure.

There are several problems with this system of equations and its relation to our theoretical model. First, Japanese wheat imports are purchased by the Japanese Food Agency (JFA), a government monopoly, which resells to wholesalers at a fixed price. The resale price is generally set annually and is typically above the import price. It is not entirely clear what mechanism the JFA uses in setting the resale price, but it would appear that their price setting behavior constitutes an implicit tariff (though it was a subsidy for two years in our sample).[10] Second, the only resale price we have available is for U.S. Western White #2 which accounts for approximately 30 percent of U.S. exports to Japan. For the U.S. regression we restrict our price data to that for Western White #2, implicitly assuming that it is an appropriate index of the price for all U.S. wheat exported to Japan as the quantity data are for all wheat. We were unable to obtain resale price data for

Canadian wheat, so we assume that the U.S. markup is applied to their exports, or

$$Jp_t^c = P_t^{jc} Jp_t^u / P_t^{j\,u}.$$

Given the problems in interpretation of the regressions and measures of the Japanese resale price, we shall assume that the relevant elasticity from the perspective of U.S. and Canadian exporters is with respect to import price in Japan. Hence, the coefficient β_1^u in the above corresponds to $-B^u$ in equation (2), β_2^c corresponds to $-B^c$ in equation (1), and β_2^u and β_1^c correspond to K. Initially, we imposed the cross-equation restriction $\beta_1^c = \beta_2^u$, but the resulting estimated coefficients violated the restriction $B^c B^u - K^2$. Since the variable Jp_t^c in the Canadian equation is measured with error we have more confidence in the U.S. equation so that we set $K = \beta_2^u$ without the cross-equation restriction. This does not violate the demand restriction.

The estimates of B^c, B^u and K are used to calculate coefficients of the inverse demand functions, (3) and (4):

$$a^u = (B^u A^c + K A^u)/(B^c B^u - K^2)$$

$$a^c = (B^c A^u + K A^c)/(B^c B^u - K^2)$$

$$b^c = B^u/(B^c B^u - K^2)$$

$$b^u = B^c/(B^c B^u - K^2)$$

and $\qquad k = K/(B^c B^u - K^2).$

From these we calculate

$$e_u^{ju} = -b^u X^u/P^{ju}$$

$$e_c^{jc} = -b^c X^c/P^{jc}$$

$$e_c^{ju} = -k X^c/P^{ju}$$

$$e_u^{jc} = -k X^u/P^{jc}.$$

Substituting the estimated elasticities into equations (15), (16) and (17) gives the estimated conjectures. Notice that the elasticity depends not only on the coefficients in the demand system but also on the level of exports and prices. This means that the inverse demand elasticities, as well as the conjectures, will vary from year to year.

It is worth noting that while availability of resale prices restricted our regression analysis, our calibration made use of data for all types of wheat imported into Japan. Our import price index was based on weights for the share of each type of wheat in Japanese imports.

Define Σ as the covariance matrix of the seemingly unrelated system (A1) and (A2). Under mild regularity conditions each of the estimated conjectures is consistent and asymptotically normal with covariance matrix given by $\Delta'\Sigma\Delta$ where Δ is the gradient of v^{cu}, v^{uc} or v^{uu} with respect to the estimated parameters of the demand system (A1) and (A2). Substituting estimated coefficients, variances and covariances of (A1) and (A2) into $\Delta'\Sigma\Delta$ gives estimated variances for the estimated conjectures.

Notes

1. See Thompson (1978) for a detailed discussion of the JFA.

APPENDIX B

Data

Data for Canadian and U.S. wheat exports to Japan are from K. Hjort, "Class and Source Substitutability in the Demand for Imported Wheat," Ph.D. dissertation, Purdue University, 1988. Import prices by type of wheat and transport costs are from International Wheat Council, *World Wheat Statistics*. Data for payments to producers in Canada are from The Canadian Wheat Board, *Annual Report*. Data for the producer price in the United States are from *Wheat Situation and Outlook*. Data on Japanese production, beginning stocks, resale price, and the Herfindahl index for U.S. exporters are from the Foreign Agricultural Service, USDA. Japanese CPI, population, GNP and relevant exchange rates are from International Monetary Fund, *International Financial Statistics*. Strike activity is from Gallagher *et al.* 1981, and updated by the International Longshoreman's and Warehousemen's Union.

Discussion

Michele M. Veeman

The integration of ideas from industrial organization theory into international trade has encompassed theoretical developments that have both positive and normative economics dimensions. The paper by Marie and Jerry Thursby relates more to the first of these dimensions. The authors postulate a theoretical model of possible price and output behavioural interelationships between specific exporters of wheat to Japan, namely the Canadian Wheat Board and the United States wheat exporting sector. Estimates of the Japanese elasticities of demand for U.S. and Canadian wheat are derived econometrically and are used to calculate conjectural variation parameters for both exporters; the standard errors of these estimates are manipulated to give estimates of the variability of the conjectures. The purpose given for the study is to ascertain the nature of exporter competition as a basis for possible future policy applications. This discussion comments first on general features of the positive and normative facets of strategic trade theory before commenting specifically on features of this particular study.

In terms of the positive economics approach, the efforts of trade theorists and empirical analysts to extend and test the theoretical framework of international trade beyond a focus on perfectly competitive markets for homogeneous products, under constant returns to scale, in the absence of externalities in consumption or production, to also consider imperfect competition, differentiated products and interdependencies in firms' (traders') actions, is to be applauded. This is particularly the case for trade in agricultural products, since world markets for these products are often relatively highly concentrated (perhaps due to economies of large scale operations but most often due to other entry barriers); farm products are variable in quality and in the nature of many of their characteristics of interest to users; and trading firms, agencies and their governments often exhibit interdependent conduct. But realism in assumptions is not necessarily the major desirable attribute of useful theoretical models. As the debate some two and a half decades ago based on Friedman and Samuelson's views on realism and theory reminded theorists, the ability to successfully explain and to predict future events is a fundamental attribute of useful economic theory (Massey 1965).

To date, it appears that movement toward integration of imperfect competition and trade theory has increased the realism of assumptions but has not yet given major improvements in prediction. To some degree this reflects characteristics of imperfect competition theory, particularly its inability to successfully predict the specific nature of interdependent actions and reactions of oligopolistic firms. Oligopolistic firms, whether trading in domestic or international markets, sometimes exhibit tendencies for cooperative strategies, such as price leadership and fellowship; alternatively, they may exhibit collusive strategies, or just follow consciously parallel actions. Periods of relative price stability that could be explained by these different types of strategies can be followed by aggressive product and price competition. Ex post, various theories and models of oligopolistic conduct can be invoked to explain behaviour. But because there is no single empirically supported model of how oligopolists behave, there is no unique ex ante explanation of the patterns of behaviour and only limited ability to predict successfully specific patterns of oligopolistic behaviour. Further, in practice, market structure and firm behaviour changes over time but convincing theoretical explanations of the dynamic nature of market structure and conduct are lacking. World trade in wheat provides an example of some of these features. For example, from the 1950s into the 1960s, Canada was viewed as a price leader in cooperation with the United States and, subsequently, with the United States and Australia (McCalla 1966; Alaouze, Watson, and Sturgess 1978). In contrast, the U.S. was the price leader in world markets for wheat in the early to mid-1980s. Periods of relative price stability in world wheat markets have subsequently been followed by intensive price warring, as in the mid-1960s and mid to late 1980s. Our ability to predict these patterns of behaviour and the changes in them has been limited.

The marriage of oligopoly theory and trade theory has generally involved fairly limited duopoly models, often in the form of two person strategic games. The number of firms is taken as exogenous. The pattern of behaviour assumed to apply involves relatively simple forms of conjectural variations compared to complex reality. Government action is also viewed as exogenous and governments are assumed to act optimally, to precommit to their actions and not to retaliate. Most of these characteristics apply to the empirical paper by Marie and Jerry Thursby. But they also apply in general to this developing field of study. These limitations do not, of course, deny the fact that theoretical and empirical progress in this area is likely to benefit both the study of trade and industrial organization. The limitations emphasize that the "new theory" is not likely to replace, but can usefully amplify, the more traditional body of international trade theory.

Advocates of strategic trade policy have argued for sectoral intervention using export subsidies, taxes or tariffs, in order to encourage home firms to achieve economies of large scale operations, to capture externalities that may be associated with some strategic sectors, or to capture rents that might otherwise accrue to foreign oligopolists or monopolists. Where imperfect competition is involved, such trade policy must be viewed as a second best alternative to the preferred first best policy, difficult to achieve in an international context, of anti-trust or competition policy. And a major problem in designing strategic trade policy to capture foreign rents is that the nature of appropriate policy is highly dependent on the structure of the trading sectors involved and also on the nature of the interdependence amongst traders, that is, on the nature of oligopolistic behaviour. As noted above, there is no generally accepted model of oligopolistic behaviour, yet appropriate strategic policy depends on the nature of oligopolistic interdependence, the number of firms, and whether or not other traders retaliate. Thus there are very few guidelines for effective strategic trade policy and many doubts. Can governments really pick "winning" agricultural sectors or firms, those that are likely to achieve major economies of large scale operations or spillover benefits to the rest of the nation and where these social benefits exceed the social costs of obtaining the funds used for such purposes? The history of domestic policy suggests, at least for agriculture, that protection often tends to go to vocal "losers" (that is, to high cost sectors) rather than to potential "winners" (low cost sectors). And if protection is given to a potentially low cost agricultural sector, will this continue to be a low cost sector? Capitalization of rents into the values of fixed assets, the transitional gains trap, leads one to doubt this. Further, even if domestic gains do exceed domestic costs, should one weight equally the gains in rent to domestic firms against the losses in increased expenditures of consumers? Finally, it is clear that effective lobbying by commodity groups and other special interest groups, rather than abstract concepts of trade and welfare theory, shapes the political economy of most trade policy for agriculture. Even some of its advocates recognize that the concepts of strategic trade policy may be misused to give intellectual respectability to protectionism pursued by organized vested interests at the expense of domestic consumers and foreign producers.

At a general level, the paper by Marie and Jerry Thursby can be commended for the efforts to test empirically the nature of conjectural variations and develop some measure of variability of these. Nonetheless, in common with much of the theory from which it evolves, the paper has a fairly narrow focus. Its analysis of the Japanese wheat import market is essentially conceptualized as a duopoly game theoretic model of two exporters, with the Canadian Wheat Board as one player and a firm representative of U.S. wheat exporters as the other. Other market players are not considered. In view of the feature that the conclusions of such trade models are highly dependent on the assumptions used, two assumptions of the model are of particular concern. First, the monopsonistic role of the Japanese Food Agency is ignored. A second major problem relates to the assumption of well behaved error terms that underlies the estimation of variability measures of the behavioural parameters, a major focus of the study.

In practice, Japanese imports of wheat involve a tendering process; weekly tenders for specific quantities of various types and qualities of wheat are specified by the Japanese Food Agency, and wheat exports, whether from Canada, the U.S., or other sources, must be made through Japanese trading firms. In view of the differences in the types of wheat that are produced in and exported from Canada and the United States, combined with the features of the tendering system, competitive interrelationships between the Canadian Wheat Board and U.S. exporters must focus on price, rather than quantity interrelationships. Assessment of the nature of competition in the Japanese wheat import market should focus on the monopsonistic importing agency. The major characteristics of the Japanese import market for wheat are not recognized by the model reported here.

The second major feature of concern with the model and resulting empirical estimates relates to the derivation of the variability measures for behavioural parameters, a procedure that involves the assumption of well behaved error terms for the estimated equations. There are a number of features relating to the econometric estimations that contribute to concern as to whether this is likely to be the case. For example, in their theoretical model the authors assume that the relevant elasticity of demand parameters relate to import prices; nonetheless, Food Agency resale prices are also included in the estimating equations. Since the authors only have these resale prices for one wheat, U.S. Western White #2, they calculate the elasticity of demand for U.S. wheat using imports and prices of this wheat only. Lacking resale prices for Canadian wheat, they synthesized these, assuming that the markup for U.S. Western White #2 wheat would also apply to Canadian wheat (despite the fact that Canadian Western Red Spring Wheat is the only Canadian wheat imported by Japan). However, from other econometric research on characteristics of wheat desired by purchasers, we know there is a preference and a premium for white wheat in Japan and a preference and a premium for Canadian wheat relative to U.S. wheat in this market, raising questions of bias in the estimates of elasticities of demand, based on the data used. The effect on error terms of including both import prices and resale prices (a synthetic price in the case of Canada) as independent variables can also be questioned. A further point of concern relates to the authors' specification of the import demand function. The theoretical development of the model involves inverse demand functions; however, the empirical estimates given here are based on direct, rather than inverse, demand functions. The impact of possible mis-specification of the estimating equations or of the evident data problems on the error terms is, unfortunately, not discussed by the authors. Indeed, full econometric results are not reported and only the standard errors of the estimated elasticities are given. Thus the reader is not able to assess whether the parameter estimates are inefficient or biased. The standard errors of the parameter estimates are certainly very high. One would like to have some indication of the stability of the elasticity estimates relative to changes in the estimation model and procedures. Finally, further interpretation of the results is desirable.

Those who grapple with the difficulties of obtaining and analyzing world trade data will sympathize with some of the problems evident here, but these problems do constrain the confidence one has in the resulting estimates. I conclude that this study is a interesting first step in trying to model oligopolistic interdependence in the Japanese wheat import market, a topic that warrants further work.

References

Alaouze, C.M., A.S. Watson, and N.H. Sturgess. 1978. Oligopoly Pricing in the World Wheat Market. *American Journal of Agricultural Economics.* 60:173-185.

Massey, Gerald J. 1965. Professor Samuelson on Theory and Realism: Comment. *American Economic Review.* 55:1155-1164.

McCalla, A.F. 1966. A Duopoly Model of World Wheat Pricing. *Journal of Farm Economics.* 48:711-727.

5

Implications of New Trade Theory for Modelling Imperfect Substitutes in Agricultural Trade

Donald MacLaren

Introduction

The modelling of international trade in agricultural products has generally treated goods as homogeneous and markets as perfectly competitive. The natural quantitative framework when transport costs are considered has been the spatial, partial equilibrium approach. Government intervention in the form of domestic agricultural sector instruments and trade policy measures has been incorporated in these models as appropriate. Recognition of barriers to imports, subsidization/taxing of exports, and particularly the existence of marketing boards and state trading organizations has led to the development of models of imperfect competition in agricultural trade, e.g., models of monopoly/monopsony and oligopoly/oligopsony. The other development, and one which has not been researched to the same extent as market structure, is trade in differentiated products. In these models differentiation has been defined more in terms of the location of production than with respect to the product's physical characteristics.

Almost a decade ago McCalla (1981) wrote that "a review of the general theory of international trade yields surprisingly few attempts to introduce concepts of imperfect competition into trade analysis" (p 16). In the intervening years the situation has changed substantially as the books by Jones and Kenen (1984), Kierzkowski (1984), Helpman and

Krugman (1985), Greenaway (1985), Greenaway and Milner (1986), Greenaway and Tharakan (1986), Krugman (1986) and Officer (1987) testify. There have been several developments in the "new" theory of international trade. A feature of most of them is that they are built on assumptions of imperfect competition and economies of scale internal to the firm. In addition they rely on special assumptions in order to achieve tractable results. For example, some models assume trade in the presence of an oligopolistic industry which produces a homogeneous good (e.g., Brander and Spencer (1985), Dixit (1984), Dixit and Grossman (1986) and Eaton and Grossman (1986)), whilst others attempt to explain intra-industry trade between two countries in models of monopolistic competition with differentiated products (e.g., Krugman (1979), Lancaster (1984) and Venables (1984)).[1]

The structure of the paper is as follows. The weakness of the spatial equilibrium model of trade in a homogeneous product is reviewed briefly, together with the theoretical assumptions which underlie a model of trade in a product which is differentiated by the location of its production (Section 2). This brief review is used as background against which to view selected recent developments in the "new" theory (Section 3). As indicated already some of these developments have assumed an oligopolistic market structure for trade in a homogeneous good, a framework which provides a rationale for potentially beneficial country-specific government policy intervention either domestically in a particular sector or through trade policy. Such models are not reviewed here. It will become apparent that the models of differentiated products which have been developed in the context of monopolistically competitive domestic markets do not provide a direct underpinning for models of agricultural trade in imperfect substitutes. Hence the structure of these models is not easily modified to reflect the realities of international trade in commodities. Nevertheless, an attempt is made to highlight specific insights which these new developments provide for building empirical models of agricultural trade (Section 4). The final Section presents some conclusions.

Conventional Approaches to Modelling Agricultural Trade

In their survey of agricultural trade modelling Thompson and Abbott (1982, p. 371) wrote "If a model is to contribute to an understanding of the interrelationships amongst countries or to be useful to decision-makers, its empirical content must reflect faithfully the essence of the structure of the markets of the regions linked through trade." Viewed against this statement, it has been demonstrated in several studies that the spatial equilibrium model fails because it cannot generate the diversity of net trade flows observed in practice nor can it produce two-way trade in apparently homogeneous products (Johnson, Grennes and Thursby 1979). Although the economic theory which underlies this model is appealing in the context of perfectly competitive domestic and international markets in which trade occurs between nations or regions as the trading units, as an adequate representation of reality it fails.

One alternative model has been advocated by Grennes, Johnson and Thursby (1978), and by Johnson, Grennes and Thursby (1979) namely, the Armington model in which goods are differentiated with respect to the location of production. They argued that importers perceive the product from domestic and from each foreign source as being different. These perceptions may rest on country-specific quality differences or on the reliability of the foreign supplier (Greenaway and Milner 1986). Both of these arguments have been advanced as explanations of the failure of the spatial equilibrium model and in support of the Armington model. But there is a third reason that has been put forward by Abbott (1979) which is that the particular policy instruments used by governments, e.g., threshold prices, break the link between world and domestic prices in such a way that there is no simple relationship between changes in domestic production and trade volumes. Spatial equilibrium models impose a one-to-one relationship. In other words changes in world prices are not necessarily transmitted into domestic price changes and hence into changes in net trade volumes.[2] This argument was expressed persuasively by Josling (1977, p. 265) who considered the interaction in international trade of domestic price policies and concluded that "the situation can be thought of as an acute 'market failure.'"

In their major study of the world wheat and coarse grain markets Grennes, Johnson and Thursby (1978) adopted the Armington model in preference to the spatial and other alternative models in order to take account of product differentiation and government policies in the grains sector.[3] However, Thompson and Abbott (1982) noted that the models were not particularly successful because of difficulties with the elasticities used which were derived more from judgement than econometric estimation. It may also have been that the Armington assumptions were too restrictive on the demand side of their models. Deardorff (1984) has questioned the Armington assumption on *a priori* grounds. He believed that while the assumption may be appropriate for products such as wine, which are highly locality specific, it is not appropriate to use location as an indicator of quality amongst suppliers. However, he was prepared to concede that the Armington model may be the best one available.

The Armington model has come under scrutiny in another way. Winters (1984) sought to test the theoretical underpinnings of the model in terms of the utility function used by Armington and the consequences for the relationships between products which stem from it. In view of the alternative utility functions which will be discussed below (Section 3), it is instructive to explore the model in more detail.

The Armington model is based on a two-stage budgeting process. Consider the case of a single country which imports a good from a number of exporting countries and which also produces a similar but not identical good domestically. In the first-stage process total expenditures on the import and domestic goods is decided on the basis of a weakly separable utility function. In the second stage the allocation of expenditure on imports from each source is then decided. Weak separability of the first-stage or upper level utility function is a necessary and sufficient condition for the second stage to be valid (Deaton and Muellbauer 1980). Hence, demand for the domestic good and the imported good are separable. Within the second-stage allocation it is assumed that the share of imports from each source can be modelled by a CES function of the form

$$m_i/m = b_i^\beta (p_i/p)^{-\beta},$$

where: m_i is the quantity of imports from country i; m is total imports of the good; b_i is the base period quantity share of country i; p_i is the import price from country i; p is the import price index for the good; and β is the constant elasticity of substitution parameter between import sources. This functional form not only forces separability amongst import sources but homotheticity on demand for imports from the various import sources (Winters 1984).

One advantage of this specification is that it is extremely parsimonious in terms of the number of parameter estimates required. In addition to an estimate of β, the only other information required is an estimate of the elasticity of demand for the imported good and trade shares data for each importer supplier. Hence, import shares can be modelled for responses to relative price or exchange rate changes for a single importing country on the basis of only two elasticity estimates and some easily obtained trade data.

However, this parsimony of parameters comes at a price in terms of the *a priori* restrictions imposed. There are a number of significant issues. Firstly, the separability assumption is probably violated if the definition of the good is too narrow (Goldstein and Khan 1985). This may well be the case in applications of the model to single agricultural commodities such as wheat, cotton and wool. Secondly, the constancy of the elasticity of substitution assumption is probably violated if the good is too broadly defined. Thirdly, because the within-good demands are homothetic, import shares are independent of income changes and only change in response to relative price changes. Moreover, the income elasticities of demand are forced to be unity for imports from all sources. This independence of import shares from income is important in the context of quality variations across import sources and is too restrictive (Winters). And fourthly, separability also restricts the substitution between import sources to be independent of the level of consumption of the domestically produced good. This too seems unreasonable in the context of a single agricultural commodity.

To test the assumptions of separability between domestic and imported products, and between different import sources, Winters estimated several AIDS models for United Kingdom imports of manufactures (SITC 5-8) from ten countries. His conclusions were: first, that demand for domestic and imported manufactures are not separable; and second, that the Armington assumptions of homotheticity, and separability between import sources, are rejected, both individually and jointly.

The Armington assumptions, despite making trade models tractable through their considerable saving in parameter estimates, appear not to produce a model appropriate for the modelling of products differentiated by origin. Do the "new" theories of differentiated products offer something better? The following section contains a review of some of them.

Selected Recent Developments in Trade Models
for Differentiated Products

Traditional models of international trade have emphasized the production side of the economy, either in terms of technology differences or differences in factor endowments, in determining comparative advantage and the gains from trade. By contrast consumer preferences have been regarded as relatively unimportant except insofar as they help to determine part of the gains from trade in homogeneous products. One aspect of some of the new theories of trade is their more detailed specification of part of the demand side of the economy. This is not to suggest that inter-country differences in preferences are the determinants of trade; rather it is to suggest that such preferences explain intra-industry trade. Traditional models explain inter-industry trade on the basis of the dissimilarities of countries but, increasingly, a larger proportion of international trade takes place between relatively similar economies, a phenomenon which cannot be explained with either Ricardian or Heckscher-Ohlin theories on their own.

Product differentiation, in the context of recently developed theories, has been defined in three different ways (Greenaway 1984). Firstly, horizontal differentiation refers to consumer preferences in which variety is valued. Within this category there are two distinct approaches: the first is referred to as the neo-Chamberlinian model in which consumer utility increases as the number of varieties available rises, i.e., the "love of variety" assumption; and the second, is referred to as the neo-Hotelling model in which each consumer has a preferred or ideal specification of each good consumed in terms of its characteristics. Secondly, vertical differentiation refers to preferences about the quality of a good, where quality reflects the absolute amounts of the characteristics contained in the product. And thirdly, technological differentiation occurs when certain attributes are incorporated in a product such that it is technically improved and, therefore, better than existing products. Although it is conceptually useful to make these distinctions, and the various trade models do so, in practice it may not be possible to separate them so clearly.

Horizontally Differentiated Products

Models of trade in differentiated products have tended to assume horizontal differentiation on the demand side and imperfect competition in the domestic industry which produces a product under increasing returns to scale. These returns to scale, internal to the firm, are assumed to be exhausted at a small level of output relative to the industry's output, and hence the market is monopolistically competitive. Differences in the consumer's sub-utility function ensures that the neo-Chamberlinian and neo-Hotelling models produce different results when trade takes place.

Neo-Chamberlinian. The nature of the utility function is as follows. Assume that preferences can be represented by a two-stage process. The upper level utility function for all consumption is given by:

$$U = U[u_1(.), u_2(.),...,u_k(.)]$$

where $U[.]$ is increasing and homothetic in its arguments (Helpman and Krugman 1985). The arguments of $U[.]$ are the sub-utility functions $u_i(.)$ which depend on the quantity, D_{ij}, of each variety j of product i being consumed. Specifically, $u_i(.)$ has been taken to be a CES function of the form

$$u_i(D_{i1}, D_{i2}, ...) \equiv \left(\sum_j D_{ij}^{\beta_i} \right)^{1/\beta_i}, \text{ where } \beta_i = (1 - 1/\sigma_i)$$

and $\sigma_i > 1$ is the elasticity of substitution between pairs of different varieties of the same product. This requirement is necessary to reflect monopolistic competition and positive marginal revenue. The sub-utility function has the following properties (Helpman and Krugman): first, every pair of varieties is equally substitutable; second, the degree of substitution does not depend upon consumption levels of any variety; and third, the function is concave and symmetrical. Moreover, it generates demand functions for individual varieties which have constant price elasticities of demand (Ethier 1987). However, this specification of preferences is inconsistent with casual observation in the sense that consumers exhibit preferences between varieties and they would not buy all varieties of a product if prices were identical.

Consumer preferences of this form have been used by Krugman (1979), Dixit and Norman (1980), Venables (1984) and Gros (1987) despite the comment above. The essence of the Krugman model is as follows. Assume that all consumers are alike with the same utility function and that the varieties of the good are produced in a single differentiated product industry. In this industry the production functions are the same for each variety and they exhibit increasing returns to scale. Therefore, each variety will be produced by only one firm and every firm will exploit its monopoly power over its segment of the market by equating marginal cost and marginal revenue. With free entry to the industry long-run profits will be zero, i.e., price equals average cost. Hence, the number of firms and varieties is determined.

When free trade opens between this economy and another, which may be similar or even identical, consumers in each country will gain because they obtain access to more varieties of the differentiated product. They will also gain if increased output lowers prices. These gains from trade are quite different from those of the traditional model which derive from terms of trade changes and increased specialization. Thus this model provides an explanation of intra-industry trade, although the direction of trade is indeterminate because it is not known which varieties will be produced in each country after trade takes place. It should also be noted that no firm is forced to leave the industry once trade occurs.[4]

Neo-Hotelling. In this approach, due to Lancaster (1980), it is assumed that the consumer prefers a particular variety of a product, the "ideal" variety. Preferences depend on the characteristics of the product which are often taken to be one-dimensional and to be represented by points on a line or points on the circumference of a circle. Moreover, it is assumed that the meaning of ideal variety is as follows. If the individual is offered a quantity of the good and is free to choose any variety, the ideal variety will be chosen independently of the quantity offered and independently of the consumption of other goods (Helpman and Krugman 1985). In addition, it is assumed that the individual will choose a variety closest to the ideal variety should the latter not be available, i.e., some substitution is possible if the ideal variety is not available. Hence, preferences are asymmetrical in contrast to those of the neo-Chamberlinian approach. The degree of substitution depends on the consumer's compensation function which is interpreted to be the quantity that is required in order to maintain a given level of utility when having to consume a variety which is not the ideal one. Therefore, the price that the consumer is willing to pay for a variety is inversely related to the distance of this variety from the ideal variety. The aggregation of individual consumer's diverse ideal varieties generates demand for variety at the aggregate level.

Lancaster (1984) used the following functional form:

$$U(q,v,y) = \left[\alpha q^{\rho}h(v)^{-\rho} + (1 - \alpha)y^{\rho}\right]^{1/\rho},$$

where q is the quantity of the differentiated good, v is the distance between the specification of this good and the consumer's ideal variety, y is the quantity of the homogeneous good which acts as the numeraire, and $1/(1 - \rho)$ is the elasticity of substitution. Other alternatives have been used. For example Eaton and Kierzkowski (1984) used the following utility function:

$$U(Y,p_i,\theta_i,Z_i) = \max [Y - p_i - |\theta_i - Z_i| , Y - \bar{p}],$$

where Z_i is the variety consumed by consumer i, p_i is the price of the differentiated good, Y is the consumer's income and θ_i is the characterization of the consumer's ideal variety.

On the production side the economy has two sectors: one producing a homogeneous good under constant returns to scale; the other is the differentiated good industry which produces under increasing returns to scale. Every variety of the differentiated good is produced with the same production function. Free entry ensures that long-run profits are zero. There is no guarantee that each consumer will find the ideal variety being produced.

When free trade is allowed, the number of varieties will increase and some consumers may then obtain exactly their ideal variety or at least a substitute closer to it than occurred in autarchy. Hence, this is one source of gains from trade: the other is that prices may fall as economies of scale are exploited.[5] If, in addition, factor endowments differ between the countries and if the differentiated product industry is relatively capital intensive, then inter- and intra-industry trade will both occur. The country that is relatively capital poor will be the exporter of the homogeneous good and the net importer of the other (Kierzkowski 1985). There is one important difference between this model and the neo-Chamberlinian model with respect to adjustment costs. In the latter model no firm is forced to leave the industry; in the former firms may leave and with them their varieties (Greenaway 1985).[6]

Vertically Differentiated Products

In this subset of models, which is much sparser than those of the previous sub-section, each product is available in a number of qualities. Associated with each quality is a different price. In the Falvey and Kierzkowski (1986) model consumers are assumed to have identical tastes, each individual prefers only one type of the differentiated product and, given relative prices, this preferred variety is determined solely by the individual's income. Since aggregate income is not uniformly distributed, there is an aggregate demand for a variety of differentiated goods. The utility function used is

$$U(x,q,y) = (x - x_0)^\alpha (q - q_0)^\beta (y - y_0)^\tau ,$$

where x, y are the amounts of the differentiated and homogeneous products respectively, q is the quality of the differentiated good, and the subscripts refer to subsistence consumption bundles. This form of the direct utility function underlies the linear expenditure system (Deaton and Muellbauer 1980).

The economy produces two goods, one homogeneous and the other differentiated. The former is produced using only labor inputs in a Ricardian production function. The differentiated good is produced in a Heckscher-Ohlin setting using one unit of labor and varying amounts of capital. The higher the quality of the product, the more capital input is required. Perfect competition is assumed in both sectors of this economy.

Trade takes place between two economies which have different factor endowments. The home country is assumed to have more efficient technology than the foreign country in the production of the homogeneous good. When free trade takes place, the price of the homogeneous product will be equalized in both countries and the wage rate in the home country will be higher than in the second. Therefore, the home country will not be competitive in the production of the lower qualities of the differentiated product. It will concentrate on producing the higher qualities which will be exported for imports of the lower qualities. Intra-industry trade is generated and its direction is determinate. This model provides a theoretical underpinning for the Armington model, insofar as it explains the location of production of the vertically differentiated product, and Falvey and Kierzkowski comment (p 159) that the Armington model may "represent a useful empirical approximation."

Empirical Issues

Deardorff (1984) noted that the models reviewed above "are too new to have led to their own empirical testing." That comment still seems to be valid today.[7] A considerable empirical literature has been generated but on the measurement of intra-industry trade rather than on the theories which have been developed to explain it.[8]

In order to test theories encompassing differentiated products, some empirical measure of differentiation is required. One measure which has been proposed, and subsequently disputed, is the Hufbauer index H. $H \equiv \sigma_{ij}/M_{ij}$, where σ_{ij} is the standard deviation of export unit values of good i being shipped to country j and M_{ij} is the unweighted mean of all unit values. It has been argued by Gray and Martin (1980) that this is not an appropriate indicator of product differentiation for a number of reasons. Firstly, the index will change in response to varying composition of export destinations through the unit values which, in addition, may reflect destination-specific market power. And secondly, it assumes that different varieties are shipped to different markets.

An alternative to the Hufbauer index is the Hedonic price index which is one option favoured by Gray and Martin (1980) and which has been used by Veeman (1987) and Feenstra (1988). The argument here is that different varieties will be reflected in different product prices. Hence, by regressing varietal prices on the characteristics of the varieties, the implied prices of the attributes can be estimated and their relative size is taken as a measure of their importance to consumers. Greenaway and Milner (1986) have also criticized this methodology on a number of grounds. Firstly, it relies on price differences to identify varietal differences. This is appropriate only in the case of vertical differentiation because in this case quality differences are often reflected in price differences. Therefore, horizontally differentiated products cannot be identified. Secondly, even if the technique is applied only to vertically differentiated products, there remains the difficulty of separating those price differences which are due to factor endowment explanations from those which are due to quality. And third, unless all relevant characteristics are included in the specification of the equation, and this assumes that characteristics are measurable, then there will be specification error and inconsistent estimates of the relationship between prices and characteristics.

Commercial Policy

In the neo-Chamberlinian model of Krugman (1979) import protection of the domestic differentiated product industry through a tariff unambiguously lowers welfare in the country imposing the protection (Greenaway 1985). The prices of imported varieties rise, their quantities fall and the number of varieties available will decline as substitution out of imported varieties takes place. It may also be the case that the prices of remaining varieties rise. Hence, the economy loses from protection and this is consistent with the conventional result for a small country.

The conclusion outlined above is sensitive to the assumptions made and Gros (1987) has arrived at a different conclusion. He has used the Krugman model to derive the optimal tariff rate for a small and a large country in the absence or presence of foreign retaliation. In the absence of retaliation there exists an optimal, positive tariff rate for the small country which forces relative prices facing consumers to equal relative social marginal costs, where the social marginal cost of domestic production is just the industry's marginal cost and that of imports is their price (exclusive of tariff). This result also holds if there is retaliation. In the large country case the optimal tariff rate forces domestic firms to price discriminate, charging a higher price in the export market than they would in the absence of intervention. If retaliation occurs, the optimal rate is reduced. Therefore, it may be concluded that the conventional result for the small country may not hold in the case of a differentiated product with neo-Chamberlinian preferences.

Further light has been shed on this issue by Flam and Helpman (1987) using a more general neo-Chamberlinian approach. They assume free entry to the differentiated product industry which, as now familiar, produces under increasing returns to scale. The other sector produces a homogeneous product under constant returns to scale. These two sectors are connected explicitly through factor markets. Their conclusions include the following: first, a tariff on the differentiated product sector is welfare improving; second, an export subsidy may increase or lower welfare; and third, an output subsidy may or may not improve welfare. Therefore, by putting more structure into the production side of the economy and linking sectors through factor markets, they show that the case for policy intervention is weakened. In particular, they suggest that if there were an additional non-competitive sector, the benefits from industrial policy would be further in doubt and, furthermore, that each case needs to be evaluated separately. In other words, they show that no general conclusions on commercial policy are forthcoming.

Neo-Hotelling models produce an even more diverse set of results. These depend upon the assumptions made about the relationships between varieties, i.e., whether the arrangement is "split" or "interleaved" on the line or circle, and whether tariffs are imposed unilaterally or reciprocally.[9] In what follows only the unilateral case will be considered. When the arrangement is a split one, tariffs result in higher import prices, possibly less variety and a loss of welfare. This result comes about because consumers whose ideal varieties are imported, cannot substitute into domestic varieties, except for those consumers at the margin of the split. Hence, they lose consumer surplus while consumers of the domestic varieties are unaffected.

In the interleaved case the outcome of a tariff is more complicated. The tariff raises the prices of imported varieties and causes substitution into adjacent domestic varieties. The prices of domestic varieties will then rise and so too will profits. Entry of new firms will occur and enlarge the number of varieties available and these are assumed to interleave with existing varieties. The outcome is that the average distance between varieties on the line or circle will decrease and this in turn will increase the price elasticity of each source's demand function. As a consequence, domestic firms and importers will be forced to price closer to their respective marginal costs. The outcome is a reduction in prices of domestic and imported varieties relative to the post-tariff, pre-entry situation (Lancaster 1984). Hence domestic consumers gain through increased variety and the lower prices of domestic varieties but they lose from the increased prices, relative to free trade levels, of imported varieties. The welfare effects may be of either sign but with Lancaster's particular parameter values, a small country gained from the imposition of the tariff. This is, of course, in contrast with the conventional result.

Greenaway and Milner (1986) were concerned about the generality of these results on a number of grounds. Firstly, the fixed costs of developing new varieties is ignored and these may nullify the net gains in consumer surplus. Secondly, the outcome depends on the continued interleaving of varieties after the tariff is imposed and they note that there is no guarantee of this. Thirdly, the assumption of free entry is important in affecting the pricing strategies of firms. And fourthly, if the improvement in welfare comes about because the tariff increases the number of varieties, this suggests that there is a domestic distortion, i.e., less than the socially optimal number of varieties in the absence of trade policy. The optimal policy instrument is then likely to be a production subsidy and not a tariff. While the results obtained by Lancaster may not be very general, at least one conclusion was reproduced by Eaton and Kierzkowski (1984) in a different neo-Hotelling model in which firms used product selection as a means of deterring new entrants. They showed that a small country can improve its welfare by imposing a tariff on imports of the differentiated product.

In the Falvey-Kierzkowski model of a vertically differentiated product the imposition of a tariff produces ambiguous welfare effects. The deadweight loss from the tariff may or may not be offset by the gains from lower import prices.[10] In addition the decrease in imports has to be compensated by a loss of exports in order to maintain balanced trade.

Implications for Modelling Agricultural Trade

The theories which have been reviewed selectively above were developed for economic structures other than those which represent the production and trade structures of agricultural products. The producers of agricultural products do not have the opportunity to undertake the type of product differentiation supposed by models of horizontal differentiation. However, it is possible to conceive of vertical differentiation being more applicable in that choices can be made to some extent about the quality of product to produce. For example, a producer may have some control over whether hard or soft wheat, long or short staple cotton, malting or feed quality barley is grown and over the micron measurement of the wool that is produced. Despite these possible quality differences, the choice of product specification is not entirely controlled by the producer, even if a niche were observed in the market. Quality often depends on the growing environment and the variety/breed characteristics over which limited control can be exercised because of location and climate.

The trade structure assumed above (Section 3) abstracts from market intermediaries. In agricultural trade these assume considerable importance. Government policy intervention has produced monopoly/monopsony marketing boards and other state trading entities. These institutional arrangements break the link between producers and consumers assumed by such theoretical models and may alter in a substantial way the type of optimal policy supported by theory. For example, Rodrik (1989) considers a country with a continuum of traders, differentiated by size, which exports a product produced by a large number of small producers. These traders are assumed to control the exporting of the product. The question which is addressed is: will these traders internalize the country's international market power? Using a partial equilibrium approach Rodrik derived firm-specific optimal tax rates. These rates turn out to be inversely related to the size of the exporting firm. Such firm-specific taxes may not be politically feasible and a uniform rate was also calculated as $(1/\varepsilon^*)(1 - H)$, where ε^* is the elasticity of foreign country demand and H is the Herfindahl index of concentration amongst exporters. This is a useful framework because it accords well with the structure of agricultural production and trade and, being partial equilibrium, can be measured relatively easily. In addition to altering price signals between producers and consumers, market intermediaries may also alter both importers' and exports' perceptions of product characteristics.

With the existence of large national or multinational commercial firms, or national monopoly/monopsony marketing boards it is more likely that horizontal differentiation of products will be possible through market intelligence activities in overseas' markets. One of the threads running through the literature on product differentiation is that firms are assumed to select a product design which is different from existing designs and so capture a part of total consumer demand. Clearly, national level organizations, whether large commercial companies or marketing boards, are better placed to do this on behalf of agricultural producers than the latter are for themselves.

In addition to the production and trade structures assumed, there is a third difference with theory. The theory outlined above has been couched in terms of consumer goods yet, with some exceptions, traded agricultural products are essentially intermediate goods. The exceptions are, for example, wine, fresh fruits and vegetables, cheese and processed foodstuffs. It may be argued, however, that processed foodstuffs belong to a different industry than do agricultural products. This point illustrates the difficulty of defining an industry which as Lloyd (1987) put succinctly is "the problem of partitioning all production and consumption activities in an economy into proper subsets which are mutually exclusive and exhaustive." Despite the very substantial literature on intra-industry trade, there still appears to be no satisfactory definition of what an industry is. Ethier (1987) has suggested that models of intra-industry trade in consumer goods can be re-interpreted as models of trade in producer goods with international economies of scale. Moreover, this re-interpretation has implications for the way in which product differentiation is viewed because the neo-Chamberlinian model is now a more appropriate characterization than the neo-Hotelling model for describing the transformation of intermediate goods into a single final product (Ethier).[11] Hence, the characteristics of agricultural products as seen from the perspective of processing firms may need to be given more attention in the specification of trade models.

Models of agricultural trade are built for a variety of purposes. These include the simulation and forecasting of trade flows under various policy scenarios, and the welfare effects of changes in various aspects of agricultural protectionism. The results from theoretical developments on commercial policy may have implications not only for the modelling of trade flows but also for the basis of policy advice given in the context of multilateral trade negotiations and protectionism. Several of the models reviewed appear to point to the optimality of trade interventions even for a small country. However, whether the well-established conventional welfare results should be abandoned still remains in doubt. It has not been established that impediments to trade in a differentiated product raise global welfare, even if they can raise welfare, relative to free trade, for the protecting country. Moreover, once the differentiated product sector has been linked with other, similar, sectors through factor markets, the case for policy intervention appears to be substantially weakened. Therefore, for agricultural trade in differentiated products free trade may still be a desirable international goal.

Perhaps the greatest insight from models of differentiated products stems from their focus on particular forms and properties of sub-utility functions and their link, albeit not one apparently acknowledged in that literature, with the Armington model which preceded it by a decade. By contrast both spatial and non-spatial price equilibrium models have focussed directly on specifying market demand functions in terms of quantities and prices. The models based on horizontal differentiation use variants of the CES functional form, while the vertically differentiated product's approach has used a form implicit in the linear expenditure model. Each of these forms has problems when confronted by data. The problems with the CES function, as used in the Armington model, have already been outlined (Section 2) but the same problems will arise with the neo-Chamberlinian model which also imposes a homothetic demand structure together with separability. Moreover, the linear expenditure system imposes severe restrictions on the properties of the model. For example, it permits only substitution relationships; it only allows goods to be inferior at the cost of non-concavity which, if allowed, results in positive own-price elasticities; and it forces approximate proportionality between price and expenditure elasticities (Deaton and Muellbauer 1980).

It would appear, therefore, that there are undesirable *a priori* restrictions imposed on demand relationships by particular functional forms. Perhaps one way forward in the modelling of agricultural trade is to make more use of the Deaton and Muellbauer Almost Ideal Demand System (AIDS) approach. Their model has three equations, the first of which is

$$w_i = \alpha_i + \Sigma_j \gamma_{ij} \log p_j + \beta_i \log (M/P),$$

where: w_i is the import share of the ith country; p_j is the price of the jth import; M is total import expenditure; P is the price index; and α_i, γ_{ij} and β_i are parameters. With this specification homotheticity is not imposed, although it can be by setting β_i equal to zero. Separability amongst import sources can be tested by setting a particular γ_{ij} equal to zero.[12]

The AIDS approach has been used with some success by de Gorter and Meilke (1987), despite specification errors in their model.[13] They analyzed the simultaneous import and export of wheat by the European Community on the assumption that imports and domestic production are not perfect substitutes. The general approach appears to be promising.

Conclusion

This paper has attempted to review those aspects of recent developments in international trade theory which are restricted to trade in differentiated products. To date there appear to be no firm conclusions emerging about the generality of the results which have been produced. Special assumptions have been made in order to investigate particular issues, e.g., how intra-industry trade arises between very similar countries, and the effects on national welfare of trade restrictions/inducements or of domestic sectoral policies. Nevertheless, despite the lack of generality, these developments have helped to focus on particular aspects of the economic structure which are overlooked in the more traditional general equilibrium models. In particular, emphasis has been placed on the role of consumer preferences for variety or for quality in the differentiated product. This focus leads to particular types of demand functions which might then be linked with an Armington type of model which retains parsimony in terms of parameter estimates but which has a more solid foundation in theory.

There is now a gap between theory and empirical models which may be difficult to fill, given the nature of the demand functions employed by theory. This lack of empirical modelling makes it difficult to assess how these new theories perform within the context of the assumed industrial structures for which they were developed, let alone assess how useful they may turn out to be for the modelling of agricultural trade.

It is apparent that traded agricultural commodities can be regarded as differentiated and imperfect substitutes. However, while some products should be viewed as final consumer products which are consistent with theoretical developments, others are better viewed as traded intermediate products for which several of these new theories would need to be re-interpreted. This distinction is important for the choice of the utility function in the horizontally differentiated models. While there remain estimation problems with AIDS models, perhaps they offer the best methodology for specifying the demand side of trade models and the relationships between imperfect substitutes in agricultural trade.

Notes

1. The impetus for the development of theories to explain intra-industry trade appears to have been partly due to the work of Grubel and Lloyd.

2. The issue of price transmission elasticities and their importance for determining the price elasticity of demand for exports facing a single country has been explored by Bredahl, Meyers and Collins (1979) and by Cronin (1979).

3. Alternative models are surveyed by Sarris (1981) and Thompson and Abbott (1982).

4. In a recent paper Markusen (1986) used a model incorporating a neo-Chamberlinian demand structure to explain the volume of intra-industry trade between East and West and factor endowments to explain the direction and volume of inter-industry trade between the capital abundant North and labor abundant South.

5. Economedes (1984) used this model to show that trade in differentiated products increases as preferences across countries become more similar. The converse also holds.

6. Dinopoulos (1988) used a model with neo-Hotelling preferences together with the 'biological' approach to national tastes and preferences to explain intra-industry trade. This appears to be a relatively new development.

7. In a recent book on empirical methods in international trade (Feenstra 1988), only one chapter addressed the empirical modelling of trade in differentiated products.

8. Greenaway and Milner (1986) surveyed the empirical issues, including documentary studies, definitional and measurement problems, and econometric results.

9. A split arrangement is one in which the domestically produced varieties are on one half of the line and the imported ones are on the other half. In the interleaved case the domestic and imported varietes may alternate along the line.

10. Bond (1988) considered optimal commercial policy for an importer of a good which was differentiated by quality.

11. The analysis of international trade in differentiated products which are intermediate rather than final goods has been explored also by Helpman and Krugman (1985).

12. This was the approach adopted by Winters and outlined briefly in Section 2.

13. See the comment by von Cramon-Taebadel (1988).

References
Abbott, P.C. 1979. The Role of Government Interference in International Commodity Trade Models. *American Journal of Agricultural Economics.* 61:135-140.

Bond, E.W. 1988. Optimal Commercial Policy with Quality-Differentiated Products. *Journal of International Economics.* 25:271-290.

Brander, J.A. and B.J. Spencer. 1985. Export Subsidies and International Share Rivalry. *Journal of International Economics.* 18:83-100.

Bredahl, M.E., W.H. Meyers, and K.J. Collins. 1979. The Elasticity of Foreign Demand for U.S. Agricultural Products: The Importance of the Price Transmission Elasticity. *American Journal of Agricultural Economics.* 61:58-63.

Cronin, M.R. 1979. Export Demand Elasticities with Less Than Perfect Markets. *Australian Journal of Agricultural Economics.* 23:69-72.

Deardorff, A.V. 1984. Testing Trade Theories and Predicting Trade Flows. Chapter 10 in R.W. Jones and P.B. Kenen, eds. *Handbook of International Economics,* Vol. I. Amsterdam: North-Holland.

Deaton, A. and J. Muellbauer. 1980. *Economics and Consumer Behavior.* Cambridge: Cambridge University Press.

de Gorter, H. and K.D. Meilke. 1987. The EEC's Wheat Price Policies and International Trade in Differentiated Products. *American Journal of Agricultural Economics.* 69:223-229.

Dinopoulos, E. 1988. A Formalization of the 'Biological' Model of Trade in Similar Products. *Journal of International Economics.* 25:95-110.

Dixit, A. 1984. International Trade Policy for Oligopolistic Industries. *Economic Journal.* 94 supplement, 1s-16s.

Dixit, A. and G.M. Grossman. 1986. Targeted Export Promotion with Several Oligopolistic Industries. *Journal of International Economics.* 21:233-49.

Dixit, A. and V. Norman. 1980. *Theory of International Trade.* Welwyn: James Nisbet & Co. Ltd.

Eaton, J. and G.M. Grossman. 1986. Optimal Trade and Industrial Policy Under Oligopoly. *Quarterly Journal of Economics.* CI:383-406.

Eaton, J. and H. Kierzkowski. 1984. Oligopolistic Competition, Product Variety, and International Trade. Chapter 5 in Kierzkowski, H. ed. *Monopolistic Dompetition and International Trade.* Oxford: Clarendon Press.

Economedes, N. 1984. Do Increases in Preference Similarity (across countries) Induce Increases in Trade? An Affirmative Example. *Journal of International Economics* 17:375-381.

Ethier, W.J. 1987. The Theory of International Trade. Chapter 1 in L.H. Officer ed. *International Economics.* Lancaster: Kluwer Academic Publishers.

Falvey, R.E. and H. Kierzkowski. 1987. Product Quality, Intra-Industry Trade and (Im)perfect Competition. Chapter 11 in H. Kierzkowski, ed. *Protection* and Competition in International Trade: Essays in Honor of W. M. Corden. Oxford: Basil Blackwell.

Feenstra, R.C. 1988. Empirical Methods for International Trade. London: MIT Press.

Flam, H and E. Helpman. 1987. Industrial Policy Under Monopolistic Competition. Journal of International Economics. 22:79-102.

Goldstein, M. and M.S. Khan. 1985. Income and Price Effects in Foreign Trade. Chapter 20 in R.W. Jones and P.B. Kenen. eds. Handbook of International Economics. Vol. II. Amsterdam: North-Holland.

Gray, H.P. and J.P. Martin. 1980. The Meaning and Measurement of Product Differentiation in International Trade. Weltwirtschaft-liches Archiv. 116:322-329.

Greenaway, D. 1984. The Measurement of Product Differentiation in Empirical Models of Trade Flows. Chapter 14 in H. Kierzkowski, ed. Protection and Competition in International Trade: Essays in Honor of W. M. Corden. Oxford: Basil Blackwell.

Greenaway, D., ed. 1985. Current Issues in International Trade: Theory and Policy. New York: St Martin's Press.

Greenaway, D. and C.R. Milner. 1986. The Economics of Intra-Industry Trade. Oxford: Basil Blackwell.

Greenaway, D. and P.K.M. Tharakan, eds. 1986. Imperfect Competition and International Trade. Brighton: Wheatsheaf Books.

Grennes, T., P.R. Johnson, and M. Thursby. 1978. The Economics of World Grain Trade. New York: Praeger Publishers.

Gros, D. 1987. A Note on the Optimal Tariff, Retaliation and the Welfare Loss from Tariff Wars in a Framework of Intra-Industry Trade. Journal of International Economics. 23:357-367.

Grubel, H.G. and P.J. Lloyd. 1975. Intra-Industry Trade: The Theory and Measurement of International Trade in Differentiated Products. London: Macmillan.

Helpman, E. and P.R. Krugman. 1985. Market Structure and Foreign Trade. London: MIT Press.

Johnson, P.R., T. Grennes, and M. Thursby. 1979. Trade Models with Differentiated Products. American Journal of Agricultural Economics. 61:120-127.

Jones, R.W. and P.B. Kenen, eds. 1984. Handbook of International Economics. Vol. I. Amsterdam: North-Holland.

Josling, T.E. 1977. Government Price Policies and the Structure of International Trade. Journal of Agricultural Economics. 28:261-276.

Kierzkowski, H., ed. 1984. Monopolistic Competition and International Trade. Oxford: Clarendon Press.

Kierzkowski, H. 1985. Models of International Trade in Differentiated Goods. Chapter 2 in D. Greenaway, ed. *Current Issues in International Trade: Theory and Policy.* New York: St Martin's Press.

Krugman, P.R. 1979. Increasing Returns, Monopolistic Competition, and International Trade. *Journal of International Economics.* 9:469-479.

———. 1986. ed. *Strategic Trade Policy and the New International Economics,* London: MIT Press.

Lancaster, K. 1980. Intra-industry Trade under Perfect Monopolistic Competition. *Journal of International Economics.* 10:151-176.

———. 1984. Protection and Product Differentiation. Chapter 9 in H. Kierzkowski, ed. *Monopolistic Competition and International Trade.* Oxford: Clarendon Press.

Lloyd, P.J. 1987. *Reflections on Intra-industry Trade Theory and Factor Proportions.* Research Paper No. 187, Department of Economics, University of Melbourne.

Markusen, J.R. 1986. Explaining the Volume of Trade: An Eclectic Approach. *American Economic Review.* 76:1002-1011.

McCalla, A.F. 1981. Structural and Power Considerations in Imperfect Agricultural Markets. Chapter 2 in A.F. McCalla and T.E. Josling, eds. *Imperfect Markets in Agricultural Trade.* Montclair, NJ: Allanheld, Osmun.

Officer, L.H. 1987. ed. *International Economics.* Lancaster: Kluwer Academic Publishers.

Rodrik, D. 1989. Optimal Trade Taxes for a Large Country with Non-atomistic Firms. *Journal of International Economics.* 26:157-167.

Sarris, A.H. 1981. Empirical Models of International Trade in Agricultural Commodities. Chapter 6 in A.F. McCalla and T.E. Josling, eds. *Imperfect Markets in Agricultural Trade.* Montclair, NJ: Allanheld, Osmun.

Thompson, R.L. and P.C. Abbott. 1982. New Developments in Agricultural Trade Analysis and Forecasting. Chapter 12 in G.C. Rausser, ed. *New Directions in Econometric Modeling and Forecasting in U.S. Agriculture.* New York: North-Holland.

Veeman, M.M. 1987. Hedonic Price Functions for Wheat in the World Market: Implications for Canadian Wheat Export Strategy. *Canadian Journal of Agricultural Economics.* 35:535-552.

Venables, A.J. 1984. Multiple Equilibria in the Theory of International Trade with Monopolistically Competitive Commodities. *Journal of International Economics.* 16:103-121.

von Cramon-Taubadel, S. 1988. The EEC's Wheat Price Policies and International Trade in Differentiated Products: Comment. *American Journal of Agricultural Economics.* 70:941-943.

Winters, L.A. 1984. Separability and the Specification of Foreign Trade Functions. *Journal of International Economics.* 17:239-263.

Discussion

Nicole Ballenger

I enjoyed Don MacLaren's paper very much. It is a good review of conventional approaches to modelling agricultural trade and also an excellent introduction, for me, to some of the new trade theory pertaining directly to trade in nonhomogeneous products. I divided my comments into three parts: first, a summary of the main themes that emerged for me in the reading of the paper; second, some comments on Don's thoughts on the applicability of the new theory to modelling agricultural trade; and third, a couple of comments on current agricultural trade policy issues as they pertain to product differentiation.

A number of key points emerged for me as I read through this paper. Most are themes that have been reiterated numerous times during this conference.

First, I was reminded of the ever apparent trade off between representation of reality and practicality or economy of method. The old standbys--the spatial equilibrium and armington models--rely on little information and strong assumptions on production structures and consumer preferences. They are said to often fail because of this. They are also parsimonious, as Don says, in their parameter and data needs. If one were to consider a product both horizontally and vertically differentiated (by source and quality, for example) and allow for imperfect substitution between all possible pairs of products, the data needs would balloon probably well beyond practicality. A recent Purdue dissertation by Kim Hjort studied elasticities of substitution for all wheat importing countries by both source (five suppliers) and class of wheat (four classes), resulting in the estimation of dozens of parameters. Other data needs would include trade shares by class of product and prices for each of these types, data likely to be extremely difficult to acquire for many agricultural goods.

Second, the paper made it clear how much more attention these models pay to the demand side of the trade equation than is seen in the "old" trade theory. This stems from the fact that this literature was designed to explain trade among similar economies, rather than among

economies with fundamentally different resource endowments or technologies. The interesting thing is that this focus generates a whole new
set of sources of gains from trade quite distinct from gains from exchange
and specialization.

Third, the characteristics of the utility functions used in the models
presented in the paper differ and these differences make a difference to
the welfare results that are derived. Thus, economists' answers are
once again bound by the assumptions that we imbed in our model
structures. The message for trade modelling is the importance of
obtaining empirical evidence of the properties of consumer demand
before specifying the forms of demand equations to be used in trade
models.

Fourth, no general conclusions on commercial policy are forthcoming
from this new set of theoretical models. This is disconcerting to those of
us who grew up believing that if there was anything we could believe
in, it was that a tariff imposed by a small country is welfare-
decreasing. It is also not very helpful to those of us who do indeed have
to make policy recommendations.

The other point that emerged from the paper is that its author didn't
find a whole lot in these new theoretical works directly applicable to
modelling trade in differentiated agricultural goods. Let me make a
few comments on this section of the paper.

Don's dissatisfaction with the contribution from these theoretical
works stems from their inappropriate representation of agricultural
production or trade structures. The production sides of these models
allow firms considerable flexibility in supplying a variety of differentiated products or products of differing qualities. Don's concern is that
agricultural producers do not have the opportunity to undertake the
type of product differentiation supposed by these models because agricultural producers often cannot offer variety (except, say, in the case of
wine) and have little control over quality. Of course this depends on
what the product is and what level of the marketing chain we are
discussing.

Products with some degree of processing are easier to differentiate, if
even through packaging alternatives, or offering alternative 'cuts' in
the case of pork and beef, parts or whole birds in the case of poultry
meat, or shelled and non-shelled products in the case of nuts. Although
weather and location are often the major determinants of quality, so
are things like pesticide and hormone use. These quality factors are of
growing importance to safety-minded consumers and within the control
of producers. Furthermore, even grain quality can be controlled once it
leaves the farm gate through inspection standards and handling

requirements. If these things are important to trade (and concerns about the quality of U.S. grain relative to that of other suppliers suggests they are), then producer and marketing groups are likely to want to strive to achieve control over them.

Further, agricultural products might be differentiated through the terms of purchase agreements. As Phil Paarlberg noted yesterday, buyers actually contract for a set of services. U.S. Government export credit guarantees are a financial service facilitating the foreign sales of many U.S. agricultural commodities. Does this service add to the quality, that is, vertically differentiate, the underlying product?

Don also notes the models abstract from market intermediaries. Thus, they don't help us deal any better than spatial or armington models with the importance of state trading in agriculture or with the possibility of monopoly power held by a small number of large traders representing a large number of small producers. Don's point is, I think, that the market power branch of the new trade theory may have more to offer than the work that has looked specifically at differentiated products, and I'm not sure I would argue with him on this point.

An interesting question, though, is how the presence of market intermediaries relates to product differentiation issues in agriculture. Don looks at this relationship briefly from the export side. He suggests that the presence of national-level marketing organizations could contribute to the ability of an agricultural industry to realize gains through horizontal differentiation because such organizations are better situated to conduct market intelligence activities than individual production units. In the United States this role does seem to be played by a number of national level commodity associations that promote the overseas sales of U.S. commodities. Clearly, the U.S. agricultural sector is aware of the importance of appealing to foreign consumer preferences and of distinguishing its product from other suppliers'. Thus, the presence of marketing intermediaries may actually help make agriculture look more like other industries with respect to product differentiation.

On the import side it is also interesting to consider how marketing agents relate to trade of differentiated products. Many importers have state trading firms that control all or most imports. To what extent do these trading entities represent underlying consumer demands for a differentiated product? To what extent do their activities reflect state preferences based, possibly, on political considerations? For example, I can easily imagine a developing country with a state trading company that prefers imports of a particular type of corn or wheat that is best suited to the production of the national bread or staple, but that also seeks diversification by source due to the government's unwillingness to be overdependent on a single supplier. How would such a situation be modelled? It seems that this is a problem that could appeal to the imperfect competition, political economy, or imperfect substitution branches of the new trade theory.

Don's final concern with the trade models he examines is their focus on final consumer goods. Agricultural goods, he argues, are more likely to be intermediate goods. I didn't find this aspect to be as disturbing as he does for a number of reasons. First, even some bulk commodities--rice and beans, for example--are essentially final goods. Second, even final consumer goods presumably pass through some additional level of the marketing chain between the port and the consumer. Thus, for all industries or goods, the models fail to recognize some stage between trade and consumption. Third, even for intermediate goods, trade presumably reflects some underlying consumer demand for the product. For example, countries that purchase only one type of wheat would do so, presumably, because people eat a bread that requires this type of wheat. The more important question would seem to be, as discussed already, whether the importing agent has a demand structure that corresponds to the representative consumer's, or whether modifications to the consumer demand model must be made in order to incorporate government behavior.

What the author liked about the new theory as it pertains to imperfect substitutes is its focus on forms and properties of utility functions. It would be nice to see this section of the paper developed further. It would be interesting to develop a taxonomy of possibilities for differentiation among agricultural goods and to consider these in terms of alternative specifications of utility functions. But fundamentally I agree with Don: the message in this body of theory is the importance of the demand side of the trade equation, something agricultural economists typically don't pay much attention to. Maybe we have spent a bit too much time worrying about how to incorporate policy levers in our trade models, and not enough time looking at the basics, the underlying properties of consumer demand.

Finally I will turn to a few comments on current agricultural trade policy questions as they pertain to trade in differentiated products. The U.S. taxpayer currently spends about $200 million a year on overseas market development programs called the Cooperator and Targeted Export Assistance (TEA) programs. These programs are designed to introduce products to overseas consumers, but they are also designed to convince consumers that U.S. products are different and better and to identify market niches. At least in the Government, we have very little understanding of the effectiveness of these programs and the effectiveness clearly relates to possibilities for and foreign consumer perceptions of product differentiation.

Bonuses provided through the U.S. Export Enhancement Program (EEP)—a program mentioned a number of times during this conference—now amount to close to a billion U.S. dollars a year. Increasingly, policy makers wish to know how to target this program so as to effectively replace EC commodity sales in overseas markets with U.S. commodity sales. This requires choosing commodity and country markets where there is a high level of substitution between the U.S. and EC products. But when we poll commodity analysts on this issue they are more likely to point to trade rigidities than to substitution possibilities. They remind us that some countries will buy only certain cuts of pork; that some seek only marbled beef, while others prefer range-fed beef; some countries want chicken parts and others want them whole. Some countries buy only Japonica rice, while others want Indica; some countries prefer soft to hard wheat, etc. Clearly, understanding imperfect substitution in commodity markets is important to understanding the possibilities for and outcomes of agricultural trade policy.

References

Hjort, Kim. 1988. Class and Source Substitutability in the Demand for Imported Wheat. Ph.D. Dissertation, Purdue University.

PART TWO

The Political Economy of Trade

6

New Developments
in the Political Economy
of Protectionism

Michael O. Moore

Trade theorists have pursued exciting new areas of research in the last decade. The most familiar innovation incorporates imperfect competition into the analysis of optimal trade policy. Some authors in this new line of research suggest that activist trade policy, so-called strategic trade policy, may improve rather than harm national welfare.[1] By calling into question old axioms about the benefits of unimpeded trade, this work has spawned vibrant discussion in academic and policy circles about optimal trade intervention.

Though the conclusions are sometimes quite divergent, both the traditional and "new" pure theory of trade investigate the economic effects of a menu of exogenously-determined trade policies. The mechanisms by which policy is made and who makes the decisions are largely ignored.

The (neoclassical) political economy of protection literature has explored these policy-making mechanisms explicitly in a concurrent burst of activity. These researchers treat commercial policy in a fundamentally different way. Policy is seen as an outcome of the interaction rational policy-makers and trade-sensitive economic groups. By considering policy as an endogenous variable instead of a piece of datum, this literature takes on the difficult task of explaining the *pattern* of protection in addition to the economic consequences of trade intervention.

The political economy of protection (PEP) and imperfect competition (IC) literature have a markedly different focus but share some common elements. Both have risen out of economists' dissatisfaction with certain aspects of the neoclassical trade paradigm. The imperfect competition literature relaxes the assumptions of atomistic competition and constant returns technology; the political economy (PEP) framework allows for utility-maximizing officials and the possibility, so evident in the "real" world, of special-interest lobbying.

The most important potential confluence of these two literatures is in the analysis of the *implementation* of strategic trade policy recommendations. Though the IC literature has shown that well-informed and benevolent governments might use activist trade policies to improve national welfare, there is no guarantee that special interest lobbying will not subvert this effort.[2] One of the original innovators in the IC literature, Paul Krugman (1987), has pointed out that ". . . an effort to pursue efficiency through [strategic trade] intervention could be captured by special interests and turned into an inefficient redistributionist program." The recent political economy literature focuses on this capture. The authors argue that profits from imperfectly competitive sectors may be dissipated through increased special-interest lobbying as industries vie for favors from an interventionist government.

The PEP literature has advanced in the last ten years from simple descriptive analyses to a richer and more sophisticated view of policy mechanisms. The authors have analyzed the political market for tariffs in models of majority voting and pressure group lobbying and have recently moved on to explaining the choice of policy instruments. The analysis has been refined to include not only rival political parties and domestic economic interests but foreign lobbyists as well.

This literature has begun to provide insights into trade policy determination but much work remains to be accomplished. Two areas for additional research involve more careful modeling of decision-making mechanisms. Authors in the endogenous protection literature, especially those who concentrate on lobbying activities, analyze single-issue political markets. This assumption masks many aspects of logrolling, vote-trading and other traits that characterize practical politics. The addition of such characteristics may have important implications for whether special interests are able to harvest the fruits of their lobbying.

The other major area for future work in the PEP literature involves the incorporation of bureaucrats and other non-elected officials into the formal models. As will be shown below, many authors argue that special-interest effectively influence trade policy decisions by vote-seeking politicians. This argument ignores the institutional arrangements currently in force in many countries, namely that civil servants largely determine short-run trade policy outcomes. This delegation of power may obviate some of the channels of influence identified in the political economy literature.

The balance of this paper provides an overview of the literature on endogenous protection.[3] The discussion is divided into four parts. The first explores the relationship of the endogenous protection literature to the pure theory of international trade. The second part briefly outlines the various traditions within the endogenous protection literature. The third section is a review of the advances in formal analytical modeling of protection developed in the last ten years. The fourth part includes some suggestions for further research.

Political Economy and the Pure Theory of Trade

The relationship of the PEP literature to traditional international trade theory can be summarized using Figure 6-1.

The pure theory is concerned with the relationships represented by link 1 in Figure 6-1. Mayer (1984) calls these sets of relations the "economic link" in commercial policy determination. The analysis is confined to the effects of alternative trade policies on production, consumption and welfare both within and between trading partners. The inquiry is conducted using various assumptions about factor mobility, domestic and international competition and technology. The maximizers in these standard models are usually limited to firms and factor owners. To the extent that the government is included, "it" is expected to maximize a social welfare function with aggregate consumption of commodities as its arguments.

Despite the familiar result that free trade is optimal for a small country in a perfectly competitive world, most nations persist in systematically and repeatedly protecting inefficient domestic industries. Indeed, some of the countries that best fit the small open economy assumptions have some of the highest levels of protection. In addition, governments often use instruments that most economists would argue are inferior to other policy options. Discriminatory duties, export subsidies, quantitative restrictions and voluntary export restraints (VERs) are becoming more common even as tariff levels retreat.

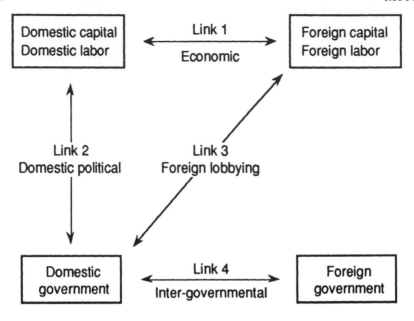

Figure 6-1. Trade Policy Determination

Two options are available to students of trade policy when confronted with these outcomes. Economists can choose to ignore them and argue that policy-makers are either irrational or ignorant. They might also simply leave the discussion to political scientists and journalists. The alternative is to apply the tools of economic analysis to these policy "aberrations" in order to explain their genesis within a framework of rational behavior. The political economy of protection literature pursues the latter option.

The recent PEP researchers take on this task by formally modeling the interaction policy-makers and trade-sensitive groups. This relationship between the government and economic interests, the so-called "political link" in trade policy, is represented by Link 2 in Figure 6-1.

In exploring this political link, the PE literature broadly addresses the following questions:

1) How is factor income affected by import competition?
2) How are policies determined and who makes the decisions?
3) What are the specific goals of policy-makers?
4) What policy instruments are available and what level of protection is chosen?

The first question is simply a restatement of questions posed in standard trade theory. The answer to this question is important in the political link because it helps identify protection-seeking coalitions. The second and third questions catalog the actors in the policy-making process and their objectives. The specific policy options open to the government and the final outcomes of the process are identified by the last question. In the terminology of game theory, these four questions identify the players, strategies, payoffs and outcomes of a policy game.

Recent work has investigated the role of *foreign* economic interests in domestic trade policy mechanisms. These extranational agents play an important role by acting as a counterweight to domestic pressure groups. This source of influence is represented schematically by link 3 in Figure 6-1 and could be called the "foreign lobbying link" in trade policy.

The direct relationship between domestic and foreign governments (link 4) might also play an important role in commercial policy determination. Political scientists such as Conybeare (1987) have considered the relationship between the governments of rival trading partners in some detail. Neoclassical economists have only rarely considered this interaction.

An appropriate modification of this schema would be to add bureaucratic component. This would recognize that the "government" is not a monolithic, single-purpose agent but instead a complex conglomeration of politicians, civil servants and political appointees, each with their own set of objectives. The bureaucracy would therefore be an intermediary between economic interests and politicians. This addition might be especially important in the short run when trade policy administration is delegated to the agencies by elected representatives. This point will be pursued in more detail below.

Traditional Models of the Political Economy of Protectionism

All political economy models are concerned with the objectives of the government and how decisions are made. Table 6-1 contains a list of the traditional PEP models and the main predictions about policy outcomes. Two distinct sets of assumptions about the political link are apparent within this literature.

Officials act to maximize some notion of national welfare in social concern models. This maximand is quite similar to the pure theory's view of the welfare-maximizing benevolent dictator. The difference lies in how social welfare is defined; instead of maximum aggregate consumption of goods, more subtle social goals are pursued.

Table 6-1. Traditional Political Economy Models

Social Concern Models

Status quo model

Conservative Welfare Function [Cordon 1974]

Governments avoid actions that lower income of any segment of society.

Adjustment Assistance Model [Cheh 1974]

Industries with high short-run adjustments will receive import relief.

Social change model

Equity concern model [Ball 1967, Constantopoulos 1974, Fielke 1974]

Concerns about social justice and low-wage workers drives trade policy.

Self-Interest Models

Pressure group model [Pincus 1975 and 1977]
Special-interest lobbying will influence trade policy.

Adding-machine model [Caves 1976]
Politicians will favor industries with large numbers of voting workers.

The assumptions of self-interest models stand in stark contrast to the altruistic social concern framework. These models contain an assumption that government officials, especially *elected* officials, maximize their own utility, even if this means a loss of overall social welfare. These models, clearly in the tradition of the public choice literature, usually have officials taking actions that increase the probability of winning an election.

Social Concern Models

There are two broad categories of social concern models within the political economy literature: (1) the status quo model and (2) the social change or equity concern model.

According to the status quo model, the government acts to protect the existing set of property rights and income distribution. Corden's (1974) "conservative social welfare function" is the classic statement of this view. Corden argues that governments tend to avoid actions that harm some members of society even if others gain. It may be true that the winners gain enough so that they could compensate the losers and still have a higher income. However, the government might recognize the practical difficulties of compensatory income transfers and opt to retain the *status quo ante* policy mix. In essence, this means giving greater weight to income losses than to income gains in the "conservative social welfare function."

A related issue is embodied in the adjustment assistance model. Cheh (1974) argues that industries with high short-run adjustment costs will be more likely to receive protection. Such industries may be ill-prepared to adjust to a new competitive environment, especially in the short-run. Cheh argues that labor force adjustment costs are particularly important when petitioning for import relief. Thus, declining industries with many older workers or with a high proportion of unskilled workers will be associated with high tariffs. Industries with highly-unionized workforces that cannot be easily dismissed may find a sympathetic ear in the trade policy circles as well.

The social change model emphasizes the use of trade policy as a means to redistribute national income or protect the poorer segments of society. Otherwise known as the equity concern model, this approach suggests that industries with low-income workers are more likely to receive protection than high-wage industries out of concern for social justice. Fielke (1976), Constantopoulos (1974) and Ball (1967) argue that low-wage industries will obtain high protection in order to redistribute income. Similarly, relatively unskilled workers are more likely to be insulated from intense foreign competition.

Self-interest Models

The other general category within the PE literature incorporates a more cynical view of the political link between economic interests and government decisions. The authors using this approach assert that policy-makers maximize their own self-interest rather than social welfare.

According to this view, governments are not driven by altruism, but instead by the desire to win elections. Ideology similarly plays a minor role; trade policy outcomes depend on the need to attract sufficient political support to retain a position of power. Two subdivisions of this approach are evident. The pressure group models emphasize narrow special interest lobbyists who try to influence vote-seeking politicians. Majority-vote models share the assumption of vote-seeking politicians but focus on the power of the general public at the polls rather than the clout of lobbyists.

These two subbranches of the literature are founded on very different assumptions about information and the transparency of governmental policy-making process. Within the pressure-group tradition, limited information among the general populace about the overall negative effects of protection allows lobbyists to influence government officials. Majority vote models presume a much less costly dissemination of information. The transparency of the political process prevents the narrow interests from benefiting at the expense of the rest of the population.

Some of the fundamental points common to pressure group models can be found in Olson's classic *The Logic of Collective Action* (1965). Olson argues that narrowly-constituted pressure groups lobby more effectively than broadly-based coalitions. Large lobbying groups will encounter difficulties in spite of their shared interests and potential voting power because of the public good nature of most governments services. No single agent can be excluded from the benefits and thus has little personal incentive to engage in applying pressure on the government. Lobbying is therefore under-supplied, at least from the standpoint of the interest group. The larger the group, the more difficult it is to identify "free riders" and the more problematic to coordinate the lobbying activities.

Pincus (1975 and 1977) applies this reasoning to tariffs. Tariffs do have many of the attributes of public goods; all industries in the importable sector benefit from the duties, regardless of whether they have lobbied for the protection or not. Pincus argues further that "the fewer the individuals who enjoy the benefits and . . . the more concentrated the benefits" the more effective is the effort to obtain protection (Pincus, 1977). A corollary is that geographically dispersed industries will find it relatively difficult to lobby. Industries whose products represent small expenditure shares in consumers budgets will also be more successful in bringing pressure to bear on the government.

Majority vote models have very different predictions about the lobbying effectiveness of large coalitions. The best known early version of a majority-vote model in the trade literature is Caves' (1976) adding machine model. This model emphasizes the potential political power of industries with large numbers of employees that are geographically dispersed. These characteristics translate into significant power at the ballot box. Caves argues in effect that the free rider problems identified by Olson has been solved. For the vote-seeking politician, granting protection to large industries that have significant workforce presence in many political constituencies increases the chance of obtaining a majority coalition in an election.

Formal Analyses of Endogenous Protection

The models of the previous section were originally advanced to help develop testable hypotheses for empirical work. They have been criticized for their neglect of formal analysis; no careful consideration was made about what conditions must hold for their predictions to be valid or whether the models were internally consistent.

A body of literature has emerged in the last decade that applies rigorous neoclassical techniques to issues of the political economy of protectionism in order to address these shortcomings. This work draws on the ideas of the earlier political literature but puts more structure on the various "links" of trade policy determination.

These formal analyses of protection are almost exclusively pressure-group models. Representative works include Brock and Magee (1978 and 1980), Findlay and Wellisz (1982), Cassing and Hillman (1985), Rodrik (1986), Magee and Young (1987) and Hillman and Ursprung (1988). Mayer (1984) is the only formal model from the majority-vote tradition considered in this review.

A summary of the assumptions and main results of these papers is contained in Table 6-2. The most important trend to note is that while the earlier work deals narrowly with the determination of the level of protection services, the analysis has been extended in recent years to selection of policy instruments.

Choosing Tariff Levels

The early formal pressure group literature is best represented by Brock and Magee (1978 and 1980). This study of endogenous tariff determination in a democracy exhibits many of the common characteristics of this literature. The relationship between lobbyists and political parties is modeled explicitly as a non-cooperative static game. The two rival political parties pick tariff policy *positions* (rather than an actual tariff level) in order to obtain campaign funds from lobbyists for use in a single-issue campaign. Free-trade and protectionist lobbyists allocate funds to the parties by equating the marginal benefit of an additional dollar of contributions with the marginal cost. Stackelberg behavior is the equilibrium concept; political parties act as leaders vis-a-vis the lobbying groups. Unlike some of the later literature, the structure of the underlying economic relationships is unspecified.

Table 6-2. Major Assumptions and Results of Formal Models

Author(s)	Economic Structure	Decision Mechanism	Agents	Choice Variable	Main Results
Mayer (1984)	Heckscher-Ohlin specific factors	Median-voter	Political parties	Tariff level	Tariff level is determined by the K/L of the median voter; voting costs allow minority industries to receive protection
Brock and Magee (1980)	Unspecified	Special-interest lobbying	Political parties and Lobbyists	Tariff level	Tariffs determined by the Cournot-Nash equilibrium level of lobbying; tariffs are not necessarily increasing in lobbying power
Findlay and Wellisz (1982)	Specific factors	Special-interest lobbying	Lobbyists	Tariff level	Welfare losses from tariffs are greater than standard theory indicates; resource-using lobbying is the reason
Magee and Young (1987)	Heckscher-Ohlin	Special-interest lobbying	Political parties, lobbyists, productive factors, voters	Tariff level	Equilibrium tariffs are *not* necessarily increasing functions of lobbying effort

Cassing and Hillman (1985)	Specific factors	Special-interest lobbying	Government, domestic monopolist	Tariff vs. quota	Tariffs will be preferred to quotas by a political-support maximizing government; domestic monopolistic market structure is crucial for the result
Rodrik (1986)	Specific factors	Special-interest lobbying	Government, domestic firms	Tariffs vs. subsidies	Welfare ranking of subsidies and tariffs may be reversed; public goods nature of tariff mean that fewer resources expended than on firm-specific subsidies
Mayer and Riezman (1987)	Heckscher-Ohlin	Unspecified	Unspecified	Tariffs vs. production cum subsidies	Regardless of policy mechanisms, tariffs will not emerge if governments can choose production taxes with subsidies to redistribute income
Hillman and Ursprung (1988)	Constant unit costs	Special-interest lobbying	Domestic and foreign lobbyists, political parties	Tariffs vs. VERs	Tariffs tend to be divisive politically; VERs are preferred to tariffs by all agents in the model so that VERs arise endogenously

The conclusions of the analysis highlight the divisive nature of much of trade policy. If both parties choose the same platform, then neither receives special-interest contributions. Similarly, the lobbyists will never contribute to both parties. The incentive structure therefore guarantees that the parties will adopt different trade policy platforms and that contributions from each lobby flows only to one political party. Perhaps the most interesting result of this model is that an increase in the influence of the tariff lobby does not necessarily lead to a higher tariff.

Findlay and Wellisz (1982) continue in the pressure group tradition, but add specific assumptions about the economic structure of the model. They focus on the additional welfare costs incurred due to tariff lobbying. This is a formalization of Tullock's (1967) point that traditional estimates of the welfare costs of tariffs are underestimated since they ignore resource-using lobbying.

These issues are addressed in a general equilibrium model with industry-specific capital as well as mobile labor. Labor can be used in productive activities or diverted to lobbying. The political process itself is a "black box"; unlike Brock and Magee, no explicit analysis of different political parties is conducted. The tariff level simply increases (decreases) monotonically as a function of pro-tariff (liberal trade) lobbying activities.

In contrast to Brock and Magee, the equilibrium is considered as the intersection of reaction functions for lobbyists rather than for political parties. The Cournot-Nash solution indicates that the equilibrium level of lobbying will cause welfare losses beyond that identified by the pure theory of trade.[4]

An important and unconventional result of this model is that welfare is not monotonically decreasing in tariffs. Intense, resource-using lobbying by both industries may nonetheless yield a relatively low tariff rate in equilibrium. The intensity of the lobbying causes significant loss of welfare. On the other hand, a high tariff that was not contentious may result in fewer welfare losses. This implies that precommitment to a certain level protection may be less damaging than a process which opens opportunities to unproductive lobbying.

Magee and Young (1987) build on the pressure-group model of Brock and Magee (1980) but include specific assumptions about economic relationships. The analysis includes utility-maximizing economic agents, lobbyists, political parties and voters within a Heckscher-Ohlin general equilibrium model. The authors explicitly consider the model to be suited for long-run analysis; factors are mobile and policymakers' preferences are completely constrained by the long-run interplay of lobbying efforts and political competition.

The lobbying equilibrium and the resultant tariff rate are determined by the interplay of lobbyists and political parties. The authors identify two major determinates of tariff changes. Unlike Brock and Magee's earlier results, a decline in the relative strength of one factor will mean that the political equilibrium will go against it. On the other hand, the "magnification paradox" indicates that increased protectionist lobbying will be accompanied by a decrease in tariff levels.

Wolfgang Mayer's (1984) insightful paper is the most important formal endogenous protection model within the majority-vote tradition. Mayer investigates trade policy determination in a general equilibrium model where decisions depend on the characteristics of the median voter. Lobbying by factor owners is not explicitly considered here. Instead, politicians adopt whatever policy commands a majority of *active* participants in the voting process.

One of the most important and innovative aspects of Mayer's model is that factor-ownership ratios vary across individuals. This is very different from the traditional view of distributional conditions. In the pressure group models above, researchers assume a one-to-one correspondence between factor-owners and factors. Thus, suppliers of labor services own no equity and capital owners provide no labor services. This assumption indicates a unique relationship between voting patterns and factor returns with significant ramifications for majority voting. One need only determine whether workers constitute a simple majority in a capital-abundant country to explain trade patterns. Mayer relaxes this artificial distinction and allows for mixed ownership of factors.

Long-run trends of the economy are analyzed using a Heckscher-Ohlin framework. Three basic conclusions are formulated: 1) an individual's optimal tariff depends on his factor endowment, 2) the larger the difference between the individual and country endowment ratio, the greater is the individual's optimal trade intervention rate, and 3) the optimal tariff is zero for individuals whose factor endowment ratios equal the national ratio. Assuming costless voting and universal suffrage, these results indicate that tariff policy outcomes depend on the capital-labor endowment mix of the median voter. Interestingly, free trade may emerge endogenously within this model if the distribution of factor ownership is symmetric. Otherwise, some form of intervention, either a tariff or export subsidy, will emerge from the voting process.

Mayer extends the analysis to a short-run model with industry-specific factors. This model is used to explain how small industries may obtain protection in a majority vote model. This result depends crucially on the presence of voting costs. Industry-specific factor losses in the importable sector may be high enough under free trade to ensure voting participation in the import-competing sector. On the other hand, losses from the imposition of a tariff may be quite diffuse; in the presence of voting, the majority of (eligible) voters may receive no *net* benefit from active participation in the trade policy vote. Tariffs are therefore an increasing function of voting costs, a result predicted in an earlier paper by Baldwin (1982).

Choosing the Policy Instrument

More recent research has shifted the focus from the determination of protectionist levels to the government's choice among protectionist instruments. Rodrik (1986) points out that the traditional literature about the positive theory of commercial policy "is better at explaining why governments may bestow special favors to import-competing sectors than at explaining why these favors tend to take the form of trade restrictions."

Research has therefore moved to investigate which policy tool will maximize the likelihood of an election victory. Only then is the particular level of protection determined. The standard modeling strategy in these investigations is to find a common level of protection for the two alternatives. The government will then choose whichever policy results in the fewest political losses for the given benchmark degree of protection.

Cassing and Hillman (1985) analyze the government's choice between a quota and a tariff in the presence of a domestic monopoly. They argue that within an endogenous policy context quotas and tariffs are not necessarily equivalent. In particular, they find that a political-support maximizing government might prefer tariffs.

In contrast to many of the models discussed above, Cassing and Hillman's analysis takes place within a partial equilibrium setting. The domestic monopoly has a fixed capital stock and employs a variable input called labor. The government chooses the type and level of trade intervention in order to maximize political-support. Any trade intervention has both political costs and benefits. The benefits from protection come as a result of increased domestic profits and subsequent higher campaign contributions. The downside is that political support from consumers will decrease with higher protection. The government must balance these two effects and pick the policy combination with the highest profits and the smallest consumer loss.

The authors find that the government can find a quota and tariff rate that result in the same higher domestic price and hence the same loss of consumer political support. However, the tariff yields higher profits and greater campaign funds from the domestic monopolist. A government would therefore choose tariffs as the interventionist tool.[5]

It is important to note that this result is driven by the assumption of domestic monopoly power. The model therefore has differential predictions for competitive and monopolistic domestic industries: noncompetitive industries should be protected by tariffs rather than quotas.

A recent article by Rodrik (1986) considers the choice between a different pair of policy instruments: tariff and subsidies. The focus is on the welfare losses associated with the two policies.

Rodrik's model is a specific-factors general equilibrium model. Like many of the other pressure-group analyses, the sector-specific capital hires the mobile factor (labor) to produce goods or lobbying services. The government's decisions are determined entirely by the lobbying resources employed by firms. Thus, the level of tariffs (or subsidy) is a monotonically increasing function of lobbying effort.

Rodrik calls into question the traditional welfare analysis conclusion that production subsidies are superior to tariffs because they generate fewer welfare losses. In particular, he finds that the ranking may be reversed if the choice is between *product*-specific tariffs and *firm*-specific subsidies.

This alternative welfare ranking depends crucially on the public goods nature of tariffs. As mentioned earlier, many authors have recognized the public goods nature of tariffs (See Baldwin 1982 and Findlay and Wellisz 1982). If import-competing firms must lobby for the implementation of a industry-wide tariff, a sub-optimal supply of lobbying will be forthcoming. In contrast, subsidies supplied at the *firm* rather than industry level are private goods. This creates a private incentive for lobbying unknown for industry-wide tariffs and consequent greater unproductive use of resources.

Wolfgang Mayer and Raymond Riezman (1987) also consider the choice of trade policy instruments in a model of variable factor ownership. They maintain a high level of generality by not specifying whether decisions are made within a median-voter or pressure group model; the choice is made "based on the decisions of individuals or groups who have the ability to shape the policy selection process."

The only assumption about their motives is that they use their trade policy choice as the sole means of redistributing income.

They argue that tariffs will not emerge as the result of endogenous policy decisions if governments can choose any combination of policy instruments. In particular, they find that a production tax cum-subsidy will make gainers and losers better off than with a tariff. They do not explicitly consider quotas in their analysis so that these results are not necessarily inconsistent with Cassing and Hillman's (1985) results that officials would choose tariffs over quotas. Instead, Mayer and Riezman address a deeper issue. What policy would be chosen if policy-makers had free rein?

The model is a Heckscher-Ohlin, general equilibrium model. Like Mayer's 1984 piece, a critical assumption is that the proportion of factor ownership differs among individuals. The authors argue that those individuals who can affect policy, by whatever means, would prefer production taxes cum subsidies since they result in higher income to themselves and fewer losses to competing interests than would any vector of trade taxes or subsidies. This result might help explain why negotiators have been successful in reducing tariffs in the postwar period while other forms of non-tariff barriers have increased in number and in scope.

The models considered so far have included only domestic agents in the analysis. This is overly restrictive since foreign economic interests have an obvious stake in domestic policy outcomes. The potential importance of this foreign influence link (link 3 of figure 6-1) has been recognized for some time (Brecher 1982) but only recently analyzed by Hillman and Ursprung (1988) and Das (1988).

Hillman and Ursprung analyze policy instrument choice with foreign interests added to the standard list of domestic political parties, domestic firms and their lobbyists. The parties' policy choice in this instance is between tariffs and VERs. VERs are preferred since they provide an external mechanism for collusion by firms. This ultimately leads to higher profits and consequent larger campaign contributions.

The analysis takes place within a partial equilibrium framework. Both domestic and foreign firms are Cournot competitors in home markets but produce imperfect substitutes. Political parties act as Stackelberg leaders vis-a-vis firms and compete for the profits earned by the less-than-perfectly competitive firms by taking trade policy positions.

Hillman and Ursprung compare the policy equilibria of tariffs and VERs in isolation. In a tariff policy equilibrium, the pro-protectionist party chooses a prohibitive tariff platform while the liberal-trade politicians will opt for a free trade policy. These platform stances put domestic and foreign firms on opposite of tariff policy debate.

On the other hand, the two political parties will announce *identical* intermediate policy platforms if export restraints are the only option. Export restraints potentially yield higher profits to *both* domestic and foreign firms and will make foreigners no worse off than under the prohibitive tariff position. In effect, the VER can truly be "voluntary" in the sense that collusive market solutions are possible due to the government restrictions. The non-divisive attributes of a VER policy means that politicians will always prefer VERs over tariffs since it increases the potential campaign contribution pool.

The analyses of Rodrik (1986), Mayer and Riezman (1987) and Hillman and Ursprung (1988) all identify conditions where politicians would prefer alternative interventionist tools to tariffs. (Cassing and Hillman (1985) is an exception to this trend.) These results may provide insight into why non-tariff barriers have become more common in recent years even as tariffs rates have declined. Political-support maximizing governments may find it preferable to dismantle tariff walls and institute barriers that yield greater political benefits anyway. As Table 6-2 shows, these results are obtained within a variety of modeling frameworks and underlying assumptions.

Future Research Areas

Recent models of endogenous protection provide important insight about the interaction of economic interests and politicians in a democracy. The characteristics of distributional coalitions that lobby for protection are identified in these formal models. The analysis is particularly helpful in understanding how vote-seeking politicians and parties arrive at trade policy positions in anticipation of an election campaign.

The models are less useful in explaining whether the platforms adopted in a campaign are implemented as public policy. Are the protectionist pronouncements translated into policy or do they become forgotten planks of an earlier election campaign? If further research indicates the latter, then worries about special-interest lobbying may be overblown. Endogenous protectionist theorists should move beyond analysis of campaign "promises" to questions of policy implementation if we are to accurately evaluate the dangers of political pressures for protection.

Two types of modeling refinements would be useful for tackling this issue. Both entail even closer attention to the institutional and legal environment in which trade policy is made. The first involves analyzing *legislative* mechanisms in addition to electoral processes. The second refinement would incorporate bureaucratic decision-making into (short-run) trade policy determination.

Including the legislative process necessarily means modeling multi- rather than single-issue decisions. Political practitioners, as opposed to political candidates, must compromise when forging legislative coalitions. This might have important consequences for a candidate who announced a pro-protectionist position to secure contributions for a relatively small constituency. The protectionist platform plank might be abandoned if a more powerful constituency can be served by trading votes. Vote-trading and log-rolling creates the possibility that the legislative outcome might be quite different from the "policy equilibria" identified by some neoclassical political economists.[6]

The other major area for future research in the political economy literature is more careful consideration of the "government." Baldwin (1985) has identified three distinct institutions whose members formulate and implement U.S. trade policy. These include the Congress, the Executive Branch and the International Trade Commission. The objectives of each may be radically different. One can imagine for example that congressional behavior corresponds to the pressure group model, presidential decisions to the foreign policy model and bureaucratic action to the adjustment assistance model.

How important might these distinctions be? In large part, it depends on the analytical time-frame. In the long run, the preferences of individual bureaucrats and politicians may be relatively unimportant. Magee and Young argue for example that "in the long run we believe that even with only two parties, party and lobby competition causes the role of policy makers to be small" (Magee and Young, 1987). In this view, the sweep of changing relative economic power will dominate not only election platforms, but accompanying legislative changes and policy outcomes as well.

In the short run however, the objectives of those who administer trade law and procedures may be of critical importance. These officials operate within a given institutional environment and implement the law that is determined in political and legislative markets. Bureaucrats in the U.S. have taken on even more importance in the post-war period. Legislatively-determined tariffs are no longer the dominant trade policy instrument. Instead, administered protection procedures, such as countervailing and antidumping duties as well as escape clause procedures are the main protectionist tools. These administrative procedures may be insulated from outside lobbying in a way that legislatively-mandated tariffs could never be.[7] At the very least, pressure would have to take on some form other than campaign contributions.

Formal theoretical analyses for this line of inquiry remain rare.[8] Two examples are Messerlin (1983) and Moore (1988). Messerlin (1983) argues that the jurisdiction of a particular bureaucracy determines the institutional attitude toward trade policy. A bureau that is a "partial dictator" for a narrow economic sector may support the factors of production it supervises. The bureaucrats may therefore advocate protection, especially non-tariff barriers that are complex, non-transparent and in need of bureaucratic administration. Indeed, Messerlin argues that this brand of bureaucrat may be more likely to supply protection than are vote-seeking politicians. Messerlin points out that bureaucracies are not uniformly interested in high import barriers. In particular, agencies that have broad responsibilities in trade policy may be significantly more supportive of liberal trade.

An agency with a broad mandate already exists in the United States. The International Trade Commission, a quasi-judicial agency, has been delegated responsibility to adjudicate trade disputes in an impartial manner. This agency is not responsible to any one industry but instead has been delegated very broad powers in trade policy. The ITC therefore is deserving of special consideration when analyzing the institutional impact on the political economy of trade.

Baldwin (1985) argues that the ITC has remained largely independent and impartial in its decisions on antidumping, countervailing duty and escape clause cases.[9] If the ITC has remained impartial in these, the most common of protectionist measures in the U.S., then (short-term) worries about political pressure for increased import barriers may indeed be exaggerated.

Moore (1988) explores whether the ITC can remain insulated from increased protectionist sentiment in a repeated game framework. The ITC, protected from direct lobbying, may be vulnerable to congressional pressure through budgetary appropriations. In the formal model, the (representative) ITC commissioner chooses the level of protection while the Congress chooses the ITC's budget. The Congress is pressured by special interests to increase import barriers. The decisions to protect, at least in the short run, lies with the ITC rather than the vote-seeking politicians. The primary research question is whether budgetary pressure will force the ITC to increase protection.

In a static version of the model, the Cournot-Nash equilibrium solution is for the ITC to ignore the congressional threats. For the repeated game, a trigger-strategy solution is derived. The effects of increased pressure for protection is ambiguous. If the ITC is the dominant player in the game, then its budget will rise, but protection may rise or fall. If the Congress dominates, they will receive higher protection but only at the cost of higher appropriations to the bureaucracy.

The conclusions of this model have two important implications. First, the importance of analyzing a policy game in a dynamic context is apparent. Without repetition one concludes that the ITC is perfectly insulated from outside pressure. This conclusion breaks down partially in a dynamic context. The other notable point is that analyzing the particular institutional arrangements has important ramifications about how protectionist pressure will be brought to bear on policy outcomes.

Students of endogenous trade policy should also investigate how firms decide which type of protection to pursue. The models of this previous section had governments choosing among policy instruments. In the short-run, with menu of policy options given, firms themselves must decide how to seek protection; their options include direct lobbying of politicians for higher trade barriers, or filing an administrative protection case such as a dumping petition or escape clause case. This decision between what Finger, Hall and Nelson (1982) call the political track and technical track for protection, respectively, has been hardly touched within a formal framework.

The endogenous protection literature must also go further in analyzing the relationship between the governments of rival trading partners. The increased emphasis on "Super 301" to pry open foreign markets through threats of retaliation is likely to become more important. Threats of punishment for "misbehavior" on the part of other government's is not addressable in the extant domestically-oriented political economy framework. The models must be modified to include the actions of foreign governments as well as allowing for collusive agreements in noncooperative games. This suggests applying the techniques of "supergames" to inter-governmental behavior.

Finally, imperfect competition and strategic trade policy should be considered within a political economy model. Of course, some form of imperfect competition lies at the heart of all lobbying models of protection; without extranormal profits, funds earmarked for campaign funds to influence policy outcomes are not available. Nonetheless, a whole range of issues that involve potential appropriation of subsidies bound for "strategically" important trade sectors should be analyzed. For example, what types of industries might be successful in "capturing" strategic trade policy mechanisms? Would firms prefer export subsidies or rather turn inward and gain more profits from a protected domestic market? Might firms use up all potential extranormal profits from strategic intervention in pursuit of limited supply of government subsidy funds? These extension will not only round out the PEP literature but may be helpful in policy discussions.

Conclusion

The purpose of this review has been to shed light onto the political economy literature's contributions to the theory of trade policy determination. The relationship of the endogenous protection literature to the pure theory has been outlined.

It seems indisputable that this literature has begun to contribute importantly to the understanding of how protection arises and why it takes a particular form. The literature has identified conditions where free trade is not a policy equilibrium, even when all agents are rational maximizers. The formal economic modeling also provides insight into how and why certain lobbying coalitions are successful in obtaining protection. The research has also helped in understanding how institutional arrangements can play a role in trade policy outcomes.

Research into the endogenous determination of protection remains in its infancy however. No economic structure is common to the models which makes comparisons of specific conclusions difficult. Specific-factors, Heckscher-Ohlin, and partial equilibrium frameworks are all represented in the literature. It seems that the specific-factors framework is the most appropriate especially when investigating special-interest lobbying; nowhere does the short-run seem to play a more important role than in politics.

This literature also largely neglects the incorporation of legal and institutional realities of trade policy into the formal analysis. In part this surely reflects the desire to keep the models manageable. Ignoring these aspects may lead to serious problems in prediction however. Consider for example the prediction that protectionist lobbying will lead governments to raise tariffs. The GATT system commits governments to a set of tariffs rate. How then do governments raise tariffs above these commitments in order to placate domestic lobbying groups? Indeed, they generally do not, instead opting for non-tariff means that involve administrative mechanisms that are quite transparent, at least in comparison to the raising of campaign funds.

Some economists may feel uneasy about addressing endogenous policy and democratic processes within an economic model. "Political support functions," "policy equilibria" and "optimal campaign contributions" are not elements of most economists' professional lexicon. Nevertheless, the gains from political economy analysis, both existing and potential, are substantial. The work may be especially useful in convincing policymakers that economists still have something to say about policy in a world where lobbyists, voters and politicians play much more than a marginal role.

Notes

1. See for example Brander and Spencer (1984) and Brander (1986) for some of the principal arguments for welfare-improving trade intervention.

2. Some critical voices concerning the ability of governments to implement the "new" strategic trade policy in a political environment include Grossman (1986) and Richardson (1986).

3. This review can touch on some of the highlights of this growing literature. Other useful surveys include Baldwin (1982 and 1989), Hillman (1989) and Magee (1984).

4. Findlay and Wellisz note that these conclusions are obtained without any presumption of revenue-seeking behavior. Welfare losses occur solely as the result of firms lobbying rather than competing from revenue generated by the tariff as in Bhagwati and Srinivasan (1980).

5. Cassing and Hillman point that the government's preference of tariffs over quotas becomes ambiguous if revenue-seeking is included in the analysis.

6. The possibility of cheating on static equilibrium "commitments" suggests further modifications on the models. The endogenous protection models are considered only in the context of non-repeated games. Since elections, lobbying and policy-making occurs over time, one should analyze these questions in a dynamic game construct. Politicians may be less likely to cheat on campaign promises in a model where they must stand again for election. This may lessen the political benefits of activities such as vote-trading.

The behavior of lobbyists may be affected by repeating the policy game as well. As Brecher (1982) points out, pressure groups clearly have incentives to collude to reduce lobbying efforts beyond the static prisoner's dilemma solution.

7. Destler (1987) argues that Congress delegated much of its authority in trade issues to noncongressional bodies was done *precisely* to insulate Congress from the lobbying activities assumed in the pressure group models above.

8. Empirical analyses of the bureaucratic role in short-run trade policy determination include Finger, Hall and Nelson (1982), Herander and Schwartz (1984), Baldwin (1985) and Moore (1989). Most find that

the agencies responsible for administered protection are by no means captive to protectionist interests.

9. Empirical analyses of ITC decisions in various types of import relief cases have yielded varying results about the agency's impartiality. Finger, Hall and Nelson (1982) find that the ITC's antidumping rulings are poorly explained by the economic variables listed in the ITC's legislative mandate. Baldwin (1985) and Herander and Schwartz (1984) find significant impartiality. Moore (1989) finds that the ITC is somewhat independent but remains vulnerable to pressure from congressional trade subcommittees.

References

Baldwin, Robert. 1982. The Political Economy of Protectionism. in J.N. Bhagwati, ed. *Import Competition and Response*. Chicago: University of Chicago Press.

—— 1985. *The Political Economy of U.S. Import Policy.* Cambridge: MIT Press.

Ball, D.S. 1967. United States Effective Tariffs and Labor's Share. *Journal of Political Economy*. 75:183-187.

Bhagwati Jagdish and T.N. Srinivasan. 1980. Revenue-Seeking: A Generalization of the Theory of Tariffs. *Journal of Political Economy*. 88:1069-87.

Brander, James. 1986. Rationales for Strategic Trade and Industrial Policy. in P. Krugman, ed. *Strategic Trade Policy and the New International Economics*. Cambridge: MIT Press.

Brander, James and B. Spencer. 1985. Export Subsidies and International Market Share Rivalry. *Journal of International Economics*. 18:83-100.

Brecher, Richard. 1982. Comment on Endogenous Tariffs, Trade Restrictions and Welfare. in J.N. Bhagwati, ed. *Import Competition and Response*. Chicago: University of Chicago Press.

Brock, William and S. Magee. 1978. The Economics of Special Interest Politics: The Case of the Tariff. *American Economic Review*. 68:246-250.

Brock, William A. and S. Magee. 1980. Tariff Formation in a Democracy. in John Black and Brian Hindley, eds. *Current Issues in Commercial Policy and Diplomacy*. New York: St. Martin's Press.

Cassing, James and A. Hillman. 1985. Political Influence Motives and the Choice Between Tariffs and Quotas. *Journal of International Economics*. 19:279-290.

Caves, Richard. 1976. Economic Models of Political Choice: Canada's Tariff Structure. *Canadian Journal of Economics*. 9:278-300.

Cheh, John. 1974. United States Concession in the Kennedy Round and Short-run Labor Adjustment Costs. *Journal of International Economics*. 4:323-340.

Constantopoulos, M. 1974. Labour Protection in Western Europe. *European Economic Review*. 5:313-318.

Conybeare, John. 1987. *Trade Wars: The Theory and Practice of International Commercial Rivalry*. New York: Columbia University Press.

Corden, W.M. 1974. *Trade Policy and Economic Welfare*. Oxford: Clarendon Press.

Das, Satya. 1989. Foreign Lobbying and the Political Economy of Protection. Unpublished manuscript, Indiana University.

Destler, I.M. 1986. *American Trade Politics: System Under Stress*. Washington: Institute for International Economics.

Fielke, Norman. 1976. The Tariff Structure for Manufacturing Industries in the United States: A Test of Some Traditional Explanations. *Columbian Journal of World Business*. 11:98-194.

Findlay, Ronald and S. Wellisz. 1982. Endogenous Tariffs, the Political Economy of Trade Restrictions, and Welfare. in J.N. Bhagwati, ed. *Import Competition and Response*. Chicago: University of Chicago Press.

Finger, J. Michael., H.K. Hall and D.R. Nelson. 1982. The Political Economy of Administered Protection. *American Economic Review*. 72:452-466.

Grossman, Gene. 1986. Strategic Export Promotion: A Critique. in P. Krugman, ed. *Strategic Trade Policy and the New International Economics*. Cambridge: MIT Press.

Herander, Mark and J.B. Schwartz. 1984. An Empirical Test of the Impact of the Threat of U.S. Trade Policy: The Case of Antidumping Duties. *Southern Economic Journal*. 51:59-79.

Hillman, Ayre. 1989. *The Political Economy of Protection*. New York: Harwood Academic Publishers, forthcoming.

Hillman, Ayre and H. Ursprung. 1988. Domestic Politics, Foreign Interests, and International Trade Policy. *American Economic Review*. 78:729-745.

Krugman, Paul. 1986. Is Free Trade Passe? *Journal of Economic Perspectives*. 2:131-144.

Magee, Stephen. 1984. Endogenous Tariff Theory: A Survey. in D. Colander, ed. *Neoclassical Political Economy: The Analysis of Rent-Seeking and DUP Activities*. New York: Ballinger Publishing Company.

Magee, Stephen and Leslie Young. 1987. Endogenous Protection in the United States, 1980-1984. in Robert Stern, ed. *U.S. Trade Policies in a Changing World Economy*. Chicago: MIT Press.

Mayer, Wolfgang. 1984. Endogenous Tariff Formation. *American Economic Review*. 74:970-985.

Mayer, Wolfgang and R. Riezman. 1987. Endogenous Choice of Trade Policy Instruments. *Journal of International Economics*. 23:377-381.

Messerlin, Patrick. 1983. *Bureaucracies and the Political Economy of Protection: Reflections of a Continental European.* World Bank Staff Working Papers, no. 568.

Moore, Michael. 1988. *U.S. Antidumping Procedure as a Repeated Game: A Theoretical and Empirical Analysis.* unpublished dissertation, University of Wisconsin, Madison.

Moore, Michael. 1989. *Rules or Politics?: An Empirical Analysis of ITC Antidumping Decisions.* unpublished manuscript, The George Washington University.

Olson, M. 1965. *The Logic of Collective Action and the Theory of Groups.* Cambridge: Harvard University Press.

Pincus, J. 1975. Pressure Groups and the Pattern of Tariffs. *Journal of Political Economy.* 83:757-778.

Pincus, J. 1977. *Pressure Groups and Politics in Antebellum Tariffs.* New York: Columbia University Press.

Richardson, J.D. 1986. The New Political Economy of Trade Policy. in P. Krugman, ed. *Strategic Trade Policy and the New International Economics.* MIT Press.

Rodrik, Dani. 1986. Tariffs, Subsidies, and Welfare with Endogenous Policy. *Journal of International Economics.* 21:285-299.

Tullock, Gordon. 1967. The Welfare Costs of Tariffs, Monopolies, and Theft. *Western Economic Review.* 5:224-232.

Discussion

Harald von Witzke

Introduction

Government market intervention in agriculture is pervasive and characterized by numerous negative economic effects. It distorts agricultural production, consumption and trade. It has adverse distributive effects and, all too often, results in burdensome budgetary expenditures.

Our profession has a long tradition of criticizing distortionary agricultural and trade policies at home and abroad, in developing as well as in developed countries. Yet our profession's impact on the course of these policies has been marginal, as policy makers tend to ignore our proposals for agricultural and trade policy reform.

It is, therefore, not surprising that agricultural economists have become increasingly interested in the determinants of agricultural trade policy decisions. The "New Political Economy" of agricultural and trade policy has established itself as a field of agricultural economics in its own right. However, it is important to keep in mind that we are just beginning to understand the forces underlying agricultural and trade policies. For our knowledge in this area to grow, it is important that we continuously keep pace with the progress made in theory and applications of government behavior in related disciplines.

Therefore, the discussant appreciates the comprehensive survey Michael Moore (1990) has given in this volume on "Recent Developments in the Political Economy of Protectionism." Following, I will deal with three aspects of Moore's paper. First, I will discuss the political economy of trade policy in the pure theory of trade. Then, I will focus on market imperfections. I will conclude with some remarks concerning the methodology of the political economy of agricultural protectionism.

Political Economy of Trade Policy in the Pure Theory of Trade

In Figure 6-1, Moore has sketched important endogenous domestic-foreign policy interactions—an aspect that is not only relevant but, as yet, not well understood.

To illustrate the relevance of endogenous international policy interdependence let us briefly look at the European Community and the United States. In both political entities the driving forces of the level of price support over time have been agricultural incomes and budgetary expenditures (von Witzke 1986, 1990). Let us assume the U.S. would unilaterally discontinue price support. This would increase world market prices and reduce budgetary expenditures under the Common Agricultural Policy of the European Community. The EC would use the budgetary savings to further increase the level of support prices there, which in turn would reduce world prices. Hence, this tendency to counteract U.S. policy liberalization would saddle U.S. agriculture with relatively high adjustment costs. Essentially the same would happen as a consequence of a growing value of the U.S. dollar or supply control programs such as the PIK, the CRP or Set-aside programs. U.S. agricultural policy would show similar adjustments to unilateral policy liberalization or to supply control in the European Community.

There appears to be general consensus among trade economists that this policy interdependence has implications for the prospects for a reform of the international agricultural trade regime. However, there is some dissent on what the specific implications for reform are. Those who are more pessimistic about the international community's ability to agree to liberalize the agricultural trade regime argue that international interdependence of national agricultural policies has made it very difficult to negotiate supranational agreements. Therefore, each country that has an interest in and is politically willing to reform its agricultural and trade policy should go ahead and unilaterally liberalize agricultural and trade policies   irrespective of what other countries do in this regard. Others, however, emphasize that it is the very fact of growing policy interdependence that allows unilateral agricultural policy reform to fail, as other countries would tend to free-ride on one country's agricultural and trade policy liberalization.

Both positions certainly have some appeal. Unfortunately, at the present state of theoretical and empirical penetration of this problem a definite answer can not be given. On the one hand we observe the attempt by the GATT signatories to negotiate a reform of the international agricultural trade regime and, with the exception of New Zealand, we do not observe unilateral agricultural policy disarmament.

On the other hand we have not yet seen a major progress in the Uruguay Round with regard to agriculture.

A related aspect deserves attention in research as well. Domestic agricultural policy decisions can significantly help or hurt farmers or consumers in foreign countries. Of course, this represents an incentive for lobbyists in other countries to exploit "Link 3" in Figure 6-1. For instance, domestic price support and/or export subsidies in a large country help consumers in foreign countries and not just the intended beneficiaries, namely domestic producers. Another example is domestic supply control in a large country which would directly or indirectly hurt foreign consumers but help foreign producers.

Therefore, both foreign consumers or producers represent potential coalition partners of domestic producer interest groups. How important the potential for foreign/domestic coalitions with regard to various policy instruments in agriculture is has not yet been systematically researched. Nor has our profession determined the existence of such international coalitions and the extent of their success.

Figure 6-1 contains an additional kind of international interdependence that has not yet attracted the attention of agricultural and trade economists. These are direct and explicit interactions between governments in the form of retaliation and/or escalation in trade policy (e.g., H. G. Johnson 1954). Economic analyses of such direct policy interactions tend to arrive at the conclusion that, on balance, the domestic economy is worse off than without such action and that the foreign country that supposedly was to be hurt economically is not much affected.

An example is the Soviet grain embargo during the Carter administration. The negative impacts on Soviet Union were marginal but U.S. agriculture may have been affected because its image as a reliable supplier on the world market had been damaged (USDA 1986).

Market Imperfections

To the extent that market imperfections affect the economic performance in agriculture both on the input and on the output side, it is appealing to speculate that they will impact the political economic markets in agriculture. However, our profession is just starting to explore the implications of imperfections on many markets. Consequently, we do not know much about their effects on the behavior of those involved in agricultural and trade policy decision making.

There is one market imperfection in agriculture, however, that agricultural economists are rather familiar with. And this is the imperfect intersectoral mobility of agricultural labor. Moore has quoted the work by Cheh (1974) who argues that industries with relatively high adjustment costs tend to enjoy relatively high rates of protection. Intuitively, this argument appears plausible as far as agriculture is concerned. Unfortunately, we do not know whether it is the "social concern" type of phenomenon that has made farm lobbies in developed countries so successful or whether it is something else. Perhaps other characteristics of agriculture are as important or even more important such as asymmetric information on the actual economic situation in agriculture, or the fact that real food prices have declined over time despite agricultural protectionism. And perhaps there are suitable theoretical approaches other than public choice theory favored by most agricultural economists.

Paarlberg (1989) has recently suggested that contract theory may help explain agricultural protectionism. Intuitively, contract theory is very appealing in this regard. Agricultural productivity growth tends to be high in developed countries. The results are declining real food prices and, given the demand elasticities, declining agricultural incomes, all other things being equal. The consequence is structural adjustment in agriculture. Consumers receive the benefits of technical progress in the form of lower food prices while agriculture bears the adjustment costs. A social contract would recognize this uneven distribution of costs and benefits of technical change and represent the basis for redistribution.

Methodology

The determinants of agricultural policy decisions have been analyzed based on a variety of methodological approaches. Rausser (1982) has suggested to distinguish between the following approaches: (i) policy reaction functions, (ii) reduced form models, and (iii) structural models of political economic markets. Adopting this categorization of our profession's approaches to the analysis of the determinants of agricultural and trade policy decisions and trying to systematize them accordingly makes it immediately obvious that virtually all our theoretical and/or empirical analyses fall into category (i) or (ii).

There have been numerous studies in which estimates of agricultural policy reaction functions are reported (e.g., Rausser, Lichtenberg and Lattimore 1982). There also is a considerable number of studies using reduced form models to implicitly or explicitly determine policy preference functions (e.g. Sarris and Freebairn 1983; Riethmuller and Roe 1986). Almost all of these models are based on the political economic calculus of agricultural policy decision makers as the suppliers of (distortionary) policy. Usually, active rent-seeking by agricultural interest groups is excluded. This is an important issue, however, as the resources used for lobbying can result in significant welfare losses in addition to those usually illustrated graphically as triangles and/or trapezoids (Tullock 1967). Moreover, there is a lack of structural models of the political economic markets in agriculture that are both theoretically tractable and empirically testable. The development of such structural models is not an easy task. If it were, somebody would have done it. The paper by de Gorter and Tsur in this volume, however, outlines the central elements of such a model. Perhaps it is the crystal-lization point for the development of structural models of political economic markets in agriculture.

Concluding Remarks

Political economic analyses traditionally have focused primarily on domestic determinants of policy decisions. International aspects of domestic policy formation are frequently neglected (Frey 1984). The paper by Moore breaks with this tradition by emphasizing numerous international aspects of trade policy decisions. This has made his paper even more relevant for IATRC members who have long been em-phasizing the importance of international dimensions of agriculture. Moore's paper has also made it clear that the state of the art with regard to the political economy of agricultural protectionism is such that any attempt to solve one research problem usually raises more new questions than it can answer.

Generating additional insights into the political economy of agricul-tural and trade policy is important both for domestic as well as for in-ternational agricultural and trade policy reform. A better understand-ing of the forces that underlie policy decisions promises to facilitate the development of policy alternatives that could not only improve social welfare but that would be politically feasible.

References

Cheh, J. 1974. United States Concession in the Kennedy Round and Short-run Labor Adjustment Cost. *Journal of International Economics*. 9:323-340.

de Gorter, H. and Y. Tsur. 1990. The Political Economy of Agricultural Policy and Trade. In this volume.

Frey, B. S. 1984. *International Political Economics*. Oxford: Basil Blackwell.

Johnson, H.G. 1954. Optimum Tariffs and Retaliation. *Review of Economic Studies*. 55:142-153.

Moore, M. 1990. New Developments in the Political Economy of Protectionism. In this volume.

Paarlberg, R.L. 1989. Is there Anything "American" about American Farm Policy. *American Journal of Agricultural Economics.*, forthcoming.

Rausser, G.C. 1982. Political Economic Markets: PERTs and PESTs in Food and Agriculture. *American Journal of Agricultural Economics.* 64:821-833.

Rausser, G.C., E. Lichtenberg and R. Lattimore. 1982. New Developments in Theory and Empirical Applications of Endogenous Government Behavior. *New Directions in Econometric Modeling and Forecasting in U.S. Agriculture.* New York: Elsevier.

Riethmuller, P. and T. Roe. 1986. Government Intervention in Commodity Markets: The Case of Japanese Rice and Wheat Policy. *Journal of Policy Modeling.* 8:327-349.

Sarris, A.H. and J. Freebairn. 1983. Endogenous Price Policies and International Wheat Prices. *American Journal of Agricultural Economics.* 65:214-224.

Tullock, G. 1967. The Welfare Costs of Tariffs, Monopolies, and Theft. *Western Economic Journal.* 5:224-232.

United States Department of Agriculture. 1986. *Embargoes, Surplus Disposal, and U.S. Agriculture.* Washington, D. C.

von Witzke, H. 1986. Endogenous Supranational Policy Decisions: The Common Agricultural Policy of the European Community. *Public Choice.* 48:157-174.

———. 1990. Determinants of the U.S. Wheat Producer Support Price: Do Presidential Elections Matter? *Public Choice.* 64:155-65 .

7

Empirical Research on the Political Economy of Trade

Edward John Ray

Introduction

Empirical research on the political economy of trade has tended to focus on two sets of issues. On the one hand, a number of studies have attempted to test hypotheses regarding the determinants of the structure of protection across industries within a given country or explain the form and level of protection provided to a given industry. Alternatively, a number of studies have attempted to clarify the political economy of protectionism by focusing on the characteristics of voters and/or their political representatives as determinants of administrative decisions on particular trade cases or votes on specific pieces of legislation. Each set of inquiries has had to wrestle with the issue of the relative strength of economic versus political factors as determinants of trade policy.

On the economic side, it is clear that special interest groups will have a tendency to work harder to obtain protection the higher are the net economic rents per member of the coalition. On the political side, the success of any coalition will depend upon the cohesiveness of the group itself, the strength and cohesiveness of opposing interests and the susceptibility of the political process to the influence of special interest groups. By their very nature, the first set of studies have tended to focus attention on the characteristics of the special interest groups while the second set of studies have focused on the characteristics of the political process.

When I made my first attempt to understand the political economy of protection, Ray (1974), it became apparent that there are two separate issues to be explained. First, what are the determinants of the level of protection that special interests within an industry seek? Assuming that the government does not simply provide prohibitive tariffs in every sector but does provide neutral or comparable support to protectionist interests across industries, what would the pattern of protection across industries look like? As it turns out, one can presume that industries are equally effective in obtaining protection and predict a pattern of tariff protection across industries in both industrialized and developing countries that has a good deal of empirical support. However, the presumption that industries have equal influence within the political system hardly explains all variations in tariff protection across industries and provides little insight into the emergence and proliferation in the forms of protection provided to different industries in recent years.

Obviously, the net rent per coalition member from protection will depend on the form of protection provided. Nontariff trade barriers, NTBs, such as quotas can be used to distribute tariff equivalent rents among members of the domestic protectionist coalition, while voluntary export restraints, VERs, can be used to bribe foreign producers and governments to limit exports. Once one begins to inquire about the form that protection takes in different industries the discussion turns to political as well as economic considerations.

Furthermore, innovations in the use of NTBs in recent years have raised the possibility that other things equal the contest for further generalized trade restraint versus trade liberalization has begun to shift in favor of protectionist interests. Therefore, for both economic and political reasons empirical work has focused increasingly on the form that protection takes in different industries and on the possibility of quantifying the effects of NTBs.

Efforts to explain both the level and form of protection provided to different industries requires some knowledge of the idiosyncrasies of particular industries and the kinds of strategies that protectionist coalitions in those industries employ in order to maximize the expected pay-off to coalition members. From an analytical stand-point, game theoretic models provide a useful framework for examining alternative coalition strategies and pay-offs. From an empirical standpoint, individual industry studies are a useful context within which to look for evidence of strategic gaming. Section II will provide an incomplete but hopefully representative sample of findings with respect to the determinants of tariffs and NTBs across industries as well as evidence of strategic behavior by protectionist interests within specific industries.

Section III will provide representative evidence regarding the determinants of voting behavior with respect to rulings in trade injury cases and with respect to specific pieces of protectionist legislation. While it is natural for the literature in section III to focus heavily on political concerns and characteristics of political institutions, economic factors are as inseparable from an explanation of voting behavior on protectionist issues as political considerations are inseparable from an explanation of the pattern of tariffs and NTBs across industries.

There has been some tendency in empirical studies of protectionism to argue that economic factors and/or characteristics of interested parties explain trade restrictions and that there is little need to dwell on purely political considerations. Efforts to attribute trade regulations to political as opposed to economic influences may have failed not because of the lack of precision in existing measures of economic and political factors but because it does not make much sense.[1] While political and economic influence within a market economy are not the same thing, they tend to be highly correlated. The political economy approach to the study of trade policy treats economic and political factors as simultaneous and interacting forces that jointly determine equilibrium conditions in the market for protection. The search for evidence of purely economic or purely political influences on the pattern of protection is unlikely to yield conclusive results.

In a recent study of U.S. export and import trade flows with 66 countries for 1978 and 1982, Summary (1989) attempted to isolate political factors that explain bilateral trade relations. She found that both exports and imports are positively related to arms transfers from NATO to U.S. trading partners (reflecting allied security interests), and to the number of foreign agents a country has registered in the United States. U.S. exports are positively related to the number of U.S. civilian government employees in a foreign country. An index of political freedom within U.S. trading partners proved to be insignificant in explaining U.S. bilateral exports and imports.

How hard is it to attach economic content to those political explanatory variables? The author herself suggests that U.S. civilian government employees abroad might help make foreigners more familiar with the United States and, therefore, more warmly disposed toward buying U.S. exports. Foreign agents registered in the United States could establish business contacts in the United States that might be useful for potential exporters back home. As Summary points out, almost half of NATO arms sales are sourced in the United States and could lead to tied sales, off-sets, of a non-military sort between the United States and trading partners. Less directly arms sales abroad may provide U.S. military goods suppliers with information about non-military export opportunities. Is the problem associated with interpreting Summary's findings mainly attributable to imprecise measures of political as opposed to economic factors? Or, is the problem simply that in this area political and economic influences are inseparable?

Section IV will conclude with a summary of the stylized facts regarding both the structure of protection within and across industries and the determinants of voting behavior both with respect to cases of claimed trade injury and to particular pieces of trade legislation. That summary of the current trade environment should be of some use to both players and spectators.

The Pattern of Protection Across and Within Industries

The primary focus of this review of empirical research on the political economy of protection will be on contributions within the last decade. There already exist several excellent surveys of empirical work on protection that include a good deal of detail with respect to material from the earlier years including Baldwin (1976b), (1984c), (1985), Hillman (1989), and Lavergne (1983). However, it will prove useful to provide some discussion of early research to indicate how the focus of empirical studies has shifted over time in response to analytical and empirical findings as well as real events in the international arena.

A number of early empirical studies including those of Stern (1963), Cheh (1976) and Stone (1978) documented the relationship between U.S. protection and the labor intensity of production of commodities in the United States. Those studies were motivated in part by the suggestion that Leontief's Paradox,[2] might be explained by distortions in trading patterns associated with tariff protection. The argument was simply that tariff protection against imports whose production was intensive in the use of labor would bias the composition of imports toward capital intensive goods. If that distortion were great enough, a capital rich country like the United States could be found to export labor intensive goods and import capital intensive goods, which represented a direct contradiction to the then accepted Heckscher-Ohlin theory of trade.

The discussion of the impact of protectionist measures on the characteristics of imports and exports invited further inquiry into the determinants of the pattern of protection across industries. From a somewhat different perspective, Balassa (1967) and others were concerned with the pattern of protection in manufacturing within the industrialized market economies and the implications of that pattern for export efforts by developing countries. Analytical work by Olson (1968), Stigler (1971), Peltzman (1976), Becker (1976, 1983), Caves (1976), Pincus (1975) and others provided a foundation for developing models that allowed for the endogenous determination of tariff protection within a country based on the characteristics of pressure groups.

Pincus (1975) argued and generally found support for the view that the likelihood that a given industry would succeed in obtaining protection from import competition would be positively related to a number of industry characteristics. For example, he found that tariff protection was positively related to the importance of the industry to the economy as measured by industry output. Protection increased with regional concentration of production, dominance of the industry by a few firms and the dispersion of sales across the country. In effect, the dominance of production by a few firms and regional concentration of production improved the chances that producers could form a cohesive coalition while the dispersion of sales implied that buyers would find it difficult to form a cohesive opposition group.

Caves (1976), identified several alternative models of tariff determination including the pressure group model. He used 1963 data on nominal and effective protection in Canada for samples of from 29 to 45 observations to test the explanatory power of the alternative models. While he found that significant coefficients tended most often to be signed in the manner predicted by his specification of the pressure group model, there were a number of variables that failed to perform as expected. He did find that geographically concentrated production, relatively large firm size and labor intensive production characteristics were positively related to tariff protection. Seller concentration tended to be negatively related to protection.

Helleiner (1977) proposed a number of changes in Caves specification of the pressure group model. Using data for nominal and effective protection rates for 1961 and 1970 for a sample of 87 manufacturing industries, he found that industries characterized by low-skill intensive methods of production receive the most tariff protection. Industries that surrendered the least amount of tariff protection in Canada between 1961 and 1970, as a consequence of the Kennedy Round of trade negotiations, tended to be industries with large numbers of small firms in which production was neither concentrated in a few firms nor heavily dependent on natural resource inputs for production purposes.

Saunders (1980) provided alternative estimates of the determinants of protection in Canada. Saunders found that seller concentration and high labor-output ratios were positively related to effective protection rates across a sample of 84 industries. Effective protection was negatively related to foreign ownership within an industry, the export share of industry sales and transport costs (presumably representing geographical dispesion of production).

The negative foreign ownership effect was interpreted as reflecting the ability of foreign producers to influence domestic policy through their foreign subsidiaries. More recently, various authors have suggested that foreign subsidiaries may have been established in the United States during the last decade in part to help parent companies fend off protectionist legislation in the United States (Hillman 1989). The negative export share effect simply reflects the fact that export interests in an industry are likely to fear retaliation by foreign countries in response to domestic protectionist measures. Therefore, export interests generally oppose tariffs and other trade barriers.

Kym Anderson (1980) found that changes in effective protection in Australia between 1973/1974 and 1977/1978 for 130 industries including 76 facing strong import pressure followed a discernible pattern. In particular, protection remained greatest in industries facing substantial increases in import pressure and rapidly rising wages during the period. Furthermore, protection was greatest for labor-intensive, low-wage, low value added industries. Presumably low-wage, labor intensive industries facing rising wage pressure and increased import competition were finding it difficult to compete head-to-head with imports and had a strong incentive to seek relief in the form of trade restrictions.

In effect, by 1980 a number of empirical studies had begun to suggest that the pattern of protection across industries was not random and that both political and economic factors seemed to influence the pattern observed. Industries that were not well suited to compete with foreign producers but had structural characteristics needed to form cohesive and effective pressure groups appeared to have both the need and the ability to seek out government protection from import competition. And, the various pieces of supporting empirical evidence had been gathered from a number of different countries and periods of time.

In his classic analysis of the pattern of protection in the United States prior to 1930, Taussig (1931) had argued quite convincingly that the pattern of tariff protection in the United States was a crazy quilt produced by politicians so anxious to pander to every special interest group demand as to have lost track of all economic sense. As partial and imperfect as the early empirical papers referred to here and others may have been, they convinced economists that further efforts to explain the contemporary pattern of protection across and within industries would not be fruitless. In fact, there is now evidence to suggest that Taussig's original characterization of the pattern of tariff protection in the United States prior to 1930 was not correct (Baack and Ray 1983).[3]

Brock and Magee (1978) developed a simple model to explain how protectionist and free trade interest groups would behave with respect to campaign contributions to politicians who propose different positions on the trade issue. The model is limited to a single election between two candidates but it explicitly incorporates the determination of protectionist positions within a model. While there may be earlier examples of papers that make the determination of tariffs endogenous in an economic model, the paper by Brock and Magee appears to have been the inspiration for many of the subsequent analytical works in this area.

Stigler (1971) developed the idea that firms in regulated industries often capture or take control of the regulatory process. In such cases, regulations that were once intended to assist in monitoring the behavior of regulated firms become tools with which the regulated firms keep out new entrants and capture economic rents for themselves. Peltzman (1976) refined and generalized Stigler's model of capture by explicitly modelling the behavior of the regulator, the government. In Peltzman's model the regulator is constrained by opposing interests from succumbing entirely to the demands of the beneficiaries of regulatory control. Ruling interests in the government face some risk of not being returned to office in future elections if they are perceived to be pawns of particular interest groups and, therefore, politicians temper their willingness to pander to protectionist demands.

Becker (1976, 1983) provided a further insight into the behavior of regulators and regulated industries. He argued that any existing regulatory scheme represents an equilibrium for the purpose of distributing rents among interested parties. Studies of regulatory schemes should take account of the fact that particular regulatory regimes emerge because they are efficient political and economic means of satisfying the conflicting demands placed on the regulatory authorities. From that perspective, changes in the pattern of protection within and across countries over time would be viewed as responses to specific changes in underlying economic and/or political conditions.

Using data for 225, 4-digit SIC manufacturing industries in the United States, Ray (1981a) examined the determinant of U.S. tariffs and NTBs in 1970. The results suggested that tariffs protected industries whose firms used low-skill intensive, labor-intensive, constant returns to scale methods of production. Such production characteristics are often associated with industries in which the United States does not compete well relative to foreign producers. NTBs were found to be associated with production of relatively homogeneous products produced using capital-intensive techniques of production in industries in which production was less concentrated. Furthermore, the simultaneous estimation of tariffs and NTBs implied that industries with historical protection in the form of tariffs were likely to receive non-tariff trade barrier protection and that industries with high tariffs were most likely to succeed in getting NTB protection too.

The fact that NTBs were associated with capital-intensive industries was interpreted to mean that recent claims for protection in the United States had come disproportionately from industries with considerable amounts of sector specific capital and that the effective control of tariffs by GATT had required new protection to be in the form of NTBs. The fact that NTBs tended to be associated with less concentrated industries was interpreted to mean that less concentrated industries that lacked the cohesiveness to get tariff protection in the past were able to get protection in the form of NTBs. We will return to that point shortly.

That industries protected with tariffs were also able to obtain NTBs was simply interpreted to mean that industries that had the ability to obtain tariff protection had a relative advantage in obtaining all forms of protection including NTBs. The association of high tariffs with NTB protection also suggested that those industries that had surrendered the least amount of tariff protection in successive GATT rounds may have actually increased their overall protection from competing imports by having obtained complementary protection in the form of NTBs.

In a separate study, Ray (1981b) attempted to explain the impact of domestic and foreign tariffs and NTBs on U.S. import and export flows in 1970 for 225, 4-digit, SIC industries and to investigate the degree to which U.S. and foreign tariffs and NTBs are simultaneously determined. The results suggested that the presence of tariffs and NTBs at home and abroad had little if any impact on U.S. trade flows and that U.S. NTBs may have been induced in part by the use of NTBs abroad. Foreign NTBs were found to be positively related to the presence and height of tariff barriers abroad. That result suggests that complementarity between tariffs and NTBs may not be a purely U.S. phenomenon.

More recently, Audretsch and Yamawaki (1988b) analyzed the effects of a number of factors on U.S.-Japanese bilateral trade flows. Using data for a number of 3-digit industries for 1977 they found that an index of NTB protection in the United States had a strong negative effect on the share of Japanese imports into the United States. At the same time, they found no evidence that nominal tariffs in the United States had an impact on Japanese import shares in U.S. markets.

Staiger, Deardorff and Stern (1988) analyzed the characteristics of bilateral trade between the United States and Japan for a sample of 22 industries that included tariff and NTB measures for 1976. Among their findings was the fact that NTBs appear more significant than tariffs for determining the factor content of traded goods for both the United States and Japan. They estimate that trade liberalization between the two countries would most likely benefit farm workers and capital in manufacturing in the United States and hurt manufacturing labor.

Audretsch and Yamawaki (1988a) analyzed the role of R&D rivalry, trade restrictions and other factors as determinants of bilateral trade between the United States and Japan. The study utilized data for 213 industries in 1977. They found that the bilateral trade balance between the United States and Japan was positively related to the ratio of an index of NTB protection in the United States divided by an index of NTB protection in Japan. So, direct measures of protection in the two countries appeared to advance U.S. rather than Japanese export interests.

However, Audretsch and Yamawaki found that R&D expenditures in Japan associated with improved product quality and with cost saving innovations reduced U.S. net exports to Japan. Furthermore, they found that Japanese industries that enjoyed access to highly subsidized depreciation allowances and presumably preferred status with MITI during the 1960s performed better in the 1970s. R&D expenditures in general were less productive in the United States than in Japan for improving the bilateral trade balance. Finally, technology transfers from the United States to Japan appear to have contributed more to Japanese bilateral trade surpluses with the United States than did technology transfers from Europe to Japan and technological developments within Japan.

The Audretsch and Yamawaki study documents the fact that policies adopted within countries that are not traditionally identified as trade restrictions can and have played a significant role in shaping bilateral trade flows. In line with much of the more recent analytical research dealing with trade strategies, we might do well to think of the international trade environment as a market setting for the exchange of goods and services that are provided by regulated industries involving various legal jurisdiction. In that context the notion of fair trade has meaning. Differences in domestic treatment of cartels, vertical and horizontal mergers, resale price maintenance schemes between wholesalers and retailers and other aspects of industrial organization across countries can imply differences in competitive advantages among producers. Trade liberalization in the fullest sense of the word would entail a rationalization of industrial regulatory schemes across countries.

Hartigan and Tower (1982) developed a set of five models of the U.S. economy that differed from each other by the degree of sector specificity attached to several skill classes of labor, physical capital, land use, and several different natural resources. They generated simulations to estimate the income redistribution effects of trade liberalization based on information about 83 sectors of the U.S. economy in 1967, including 63 traded goods and 20 nontraded goods. If one presumes that all factors are mobile in the long-run, the alternative models really capture short-run and long-run effects of trade regime changes on various factor shares.

Hartigan and Tower find that owners of land and extractive resources, scientists and professionals are made worse off either by unilateral or multilateral trade liberalization. Except in the short-run, capital owners benefit from mutual trade liberalization. On the other hand, reciprocal movements toward free trade hurt farmers and craftsmen under all time horizons. And, less skilled labor would lose significantly from unilateral or multilateral trade liberalization in the short-run and gain in the long-run. They suggest that myopic behavior on the part of labor union leaders and politicians may be associated with protection of low-skill labor in order to avoid substantial short-run losses despite anticipated long-run gains.

Robert Baldwin (1982) speculated about several factors that might influence the pattern of protection across industries that quickly found empirical support. First he noted that efforts to secure protection in an industry that could capture super-normal profits would be plagued by the possibility that success would bring new entrants into the industry who would drive pure profits back down to zero. Therefore, he reasoned, concerted efforts to gain protection might be more likely in declining industries whose success in obtaining protection would merely keep existing firms in business and not attract new entrants.

Second, he suggested that there may be a conservative social welfare bias in the provision of protection.[4] Protection might be used to minimize redistributions of wealth induced by changing international economic and political conditions. That suggestion implies, as did the earlier idea, that in the face of changing world conditions, protection will go to declining industries rather than expanding industries. Alternatively, to the extent that market conditions remain fairly stable over time, the pattern of protection observed across industries would also remain stable over time.

Finally, Baldwin suggested that the absence of private insurance to protect industries from losses associated with diminished competitiveness might lead society to provide a kind of safety net for industry.[5] The systematic use of trade restrictions could moderate the rate of decline of domestic production in failing industries.

In that same year, Hillman (1982) developed a simple Stigler-Peltzman type model in which the government could be induced to provide protection for declining industries. The most interesting implication of the model was the fact that government support falls short of fully offsetting industry losses due to foreign competition. Therefore, government intervention slows but does not prevent the ultimate decline of non-competitive industries.

Lavergne (1983) studied the political economy of U.S. tariff protection at several points in time including analyses of post-Kennedy Round and post-Tokyo Round rates of tariff protection. He found that measures of comparative disadvantage have the expected positive association with U.S. tariffs. He finds no link between tariffs and slow growth industries and finds that the most significant variable explaining tariff protection even after the Tokyo Round i.e. including, therefore, current tariffs in the United States, is the pattern of tariff protection embodied in the Smoot-Hawley tariff of 1930. One interpretation of those results would be that market conditions facing historically protected industries in the United States, including their lack of competitiveness, have not changed substantially in the last 60 years. Therefore, the relative protection afforded to different industries has not changed. Such an argument would follow from the conservative social welfare hypothesis described earlier.

Alternatively, one might argue that most new protection since World War II has come in the form of NTBs. And, we already know that NTBs have been used to supplement high tariff protection for some industries and to provide new protection for capital-intensive, less concentrated industries that have not enjoyed protection in the form of tariffs in the past. In fact, as Baldwin hypothesized earlier with respect to protection in general, Lavergne found that non-tariff trade restrictions were more prevalent in declining industries. Again, GATT constraints on the use of tariffs to protect "newly" declining industries in the post-war period meant that NTBs had to be used to do the job.

Marvel and Ray (1983) argued that changes in protection like those observed during the Kennedy Round should reflect changes in underlying political and economic conditions in the spirit of the Stigler-

Peltzman model. Furthermore, they argued that GATT Rounds embodied efforts to liberalize trade but they also provided the opportunity to restructure the pattern of protection across industries. In particular, Marvel and Ray argued that if tariff rounds served the general purpose of re-regulating trade among nations, some industries might actually gain protection at the same time that the general thrust of the negotiations was to reduce protection.

They found that post-Kennedy Round tariffs systematically differed from 1965 tariffs in predictable fashion. The 1970 rates were higher for industries with relatively concentrated production, for industries that had experienced slow growth in a recent ten year period, and for industries using low-skill intensive methods of production. They also found that post-Kennedy Round tariff rates were higher in industries producing goods for final sale to consumers, reflecting the fact that a lack of buyer concentration prevented consumers from forming a lobby to offset the producers' coalition.

Marvel and Ray also analyzed the structure of NTB protection across industries in the United States. They found that NTBs were associated with industries that had relatively high tariffs before the Kennedy Round and with industries that experienced the smallest cuts in tariffs during the Kennedy Round. That last observation represented further confirmation of the fact that NTBs were used not simply to substitute for tariff protection lost during the Kennedy Round but also to provide complementary protection, Ray (1981a), to industries that had sacrificed the least amount of protection during the Kennedy Round.

Additional findings included a positive relationship between the share of sales to final consumers in an industry and NTB protection and a clear association of NTBs with less concentrated production. The relationship between low concentration on the production side and the use of NTBs had been found earlier by Ray (1981a). Marvel and Ray argued that the greater flexibility one has with NTBs relative to tariffs to reward participants in a protectionist coalition means that less concentrated industries unable to forge effective coalitions for tariff protection because of free-rider problems could obtain NTB protection.[6] The relative efficiency of NTBs in coping with the free-rider problem raises the possibility that other things equal, innovations in the use of NTBs will make protectionist victories more likely in the future than would have been the case when trade restrictions were defined predominantly in terms of tariffs.

Baldwin (1984) has provided a systematic review of protectionist policies in the United States throughout the post-war period. He recounts the fact that after surrendering much of the discretionary authority to make trade policy to the President during the depression, Congress has systematically reasserted its constitutional right to shape international trade agreements by the United States. With each successive bill authorizing U.S. participation in GATT negotiations to liberalize trade since World War II, Congress tightened restrictions on Presidential discretion regarding multilateral agreements. For example, while Congress authorized the President to seek multilateral agreements with respect to nontariff trade restrictions under the Trade Act of 1974, it required that any agreements that were reached would have to be approved by a majority vote in both Houses of Congress. That stipulation also was included in the Trade Agreement Act of 1979. Congress had imposed no such conditions on earlier negotiating efforts.

No doubt the limitations Congress imposed on the President with respect to negotiations on NTBs contributed to the lack of progress achieved during the Tokyo Round with respect to controlling the use of NTBs. Those bills along with the Omnibus Trade and Competitiveness Act of 1988 also diminish the likelihood that substantial gains in trade liberalization will result from the current Uruguay Round.

When the Kennedy Round negotiations were undertaken a number of developing countries complained that the pattern of liberalization achieved provided them with little gain in terms of their access to industrialized country markets for their manufactured exports. Ray and Marvel (1984) attempted to assess the impact of the Kennedy Round tariff cuts on the likelihood that developing countries could substantially increase their exports of manufactures to developed countries including the United States, Canada, the EC and Japan. Using 1975 data for 328, 4-digit, SIC manufacturing industries, they analyzed the patterns of post-Kennedy Round nominal and effective protection rates and NTB use in each of the four industrial areas.

The major focus of the study was on the impact of tariffs and NTBs in the industrialized markets on imports of consumer goods, textiles and processed agricultural products. Those product groups were generally believed to be potential fast growth prospects for manufactured exports from developing countries to the developed countries. The study indicated that post-Kennedy Round tariff protection went to producers of consumer goods and textiles and to industries using low-skill intensive techniques of production. U.S. NTBs were concentrated in industries that had experienced the least loss of tariff protection as a result of the Kennedy Round cuts in general and to producers of consumer goods, textiles, and processed agricultural products.

Japanese tariffs were associated with production of processed foods and consumer goods while NTBs were systematically associated with processed foods and industries employing R&D intensive methods of production. Ray and Marvel conjectured that Japanese protection may have contributed to the growth of high-tech industries in Japan.

Post-Kennedy Round tariffs were found to be associated with consumer goods, processed agricultural products and textiles in the EC, while NTBs were associated with consumer goods and processed agricultural products. Canadian tariffs were also associated with consumer goods, textiles and processed agricultural products and more generally with industries using low-skill intensive methods of production. Canadian NTBs supplemented tariff protection in textiles and processed foods.

In more recent studies (Clark 1987 and Ray 1987, 1989a, 1989b) evidence has been collected that suggests that protectionist interests are not easily brushed aside. The United States adopted a Generalized System of Preferences (GSP) in 1975 with the expressed purpose of providing developing countries with compensatory, duty-free access to U.S. markets for their exports of manufactured goods that had been largely ignored in the Kennedy Round tariff cuts. But, the evidence suggests that the GSP did not induce increases in exports of consumer goods, textiles, or processed agricultural products from developing countries to the United States.

In fact, the GSP appears to have provided perverse incentives for developing countries to expand their exports of manufactures to the United States in product categories in which their long-term export possibilities are least promising. Ray (1987) explained that result as a consequence of the fact that the GSP allowed developing countries to gain duty-free access to U.S. markets in which they were least threatening to U.S. business interests. To the extent that developing countries responded to the incentives implicit in the program, they diverted their export activities away from their most competitive manufacturing export industries.

The Caribbean Basin Initiative (CBI), adopted in 1983 with the expressed purpose of providing Caribbean nations with duty-free access to U.S. markets, has some of the same characteristics as the original GSP. Given the political rush with which the CBI was adopted following the invasion of Grenada, protectionist interests appear to have been less successful in undermining the value of the CBI compared to their ability to sabotage the expressed intent of the GSP program. Therefore, the perverse export inducement effects of the GSP legislation are not as apparent in the CBI.

Unfortunately, the revision of the GSP in late 1984 did little to redress the perverse export incentive effects of the original GSP. Ray

(1989a, 1989b) finds evidence to support the hypothesis that the revised GSP is less promising than the original legislation as an instrument to assist developing countries in their efforts to export manufactured goods to the United States. And, there is evidence to suggest that the revised GSP will not provide any advantages to major world debtors seeking to expand manufactured exports to the United States in order to earn the hard currency needed to meet their external debt payment requirements.

Chow and Kellman (1988) investigated whether or not the apparent bias in trade liberalization away from manufactured exports of greatest interest to developing countries represented a form of bias against those countries. They concluded that there is no evidence of a conspiracy by industrial countries against manufactured exports from less developed countries (LDCs). They argue that LDCs face greater import barriers in the United States and other industrial countries for their exports in the post-Kennedy Round and post-Tokyo Round periods because the developing countries have gained a comparative advantage in exporting certain product categories in the last 20 years in which protectionist forces within the industrialized countries are most effective in obtaining relief from foreign competition.

Baldwin (1985) provides an excellent review of much of the theoretical and empirical research on the political economy of U.S. protectionist policies available at the time. Baldwin provided fresh evidence regarding the determinants of U.S. tariffs and NTBs.[7] One novel piece of empirical work included using information on 262 industries, 119 of which were granted exceptions from the Kennedy Round tariff cuts in the United States, to predict which industries were granted exceptions. The results indicated that exceptions tended to be granted to high employment industries using low-skill and labor-intensive methods of production. Given the United States' lack of international competitiveness in the production of such products it is not surprising that industries granted an exception also faced high import penetration ratios in their domestic markets.

Among a number of other interesting findings were a few that are familiar from earlier studies. For example, Baldwin found that industries with high tariffs also benefitted from NTB protection. While industries with tariff protection tended to employ labor intensive methods of production, industries that enjoyed the benefit of NTB protection tended to employ more capital-intensive production techniques.

Baldwin also estimated a two-stage model for the demand and supply of protection in the United States in an effort to explain tariff reductions during the Tokyo Round. The estimated results are not highly significant but they do suggest that the demand for protection is greatest in industries employing labor-intensive techniques of production that have historically enjoyed substantial tariff protection and have suffered substantial losses of domestic sales to foreign producers. The supply of protection appears to be responsive to the intensity of industry demand for government assistance and to industries whose production methods are low-skill and labor intensive in general.

Baldwin interprets his findings as consistent with some aspects of several models or hypotheses about the characteristics of protected industries. The inadequacy of the so-called "Pressure Group" model to explain patterns of protection may be attributable in part to the fact that in most representations there is no attention paid to market conditions and the characteristics of competitive and non-competitive industries. The pressure group model simply identifies characteristics of industries that would have the greatest likelihood of forming effective coalitions to obtain relief from foreign competition.

Clearly, industries differ in the degree to which they are desperate for government assistance. One would expect to find some industries that are able to put together successful protectionist coalitions because what they lack in terms of organizational advantages relative to other coalitions is more than compensated for by the desperation of their economic situation. With that view in mind, there is nothing surprising about the finding that industries characterized by low-skill, labor intensive methods of production in the United States are generally protected by trade restrictions. Such industries are not likely to be able to compete against foreign firms for domestic sales. Producers and workers in such industries may face greater costs of forming effective coalitions to get government assistance than special interest groups in other industries but they are among the most desperate for that help. A model of pressure group competition for government protection that incorporates differences in market conditions and competitiveness across industries is more consistent with empirical findings than is generally recognized.

Using information from an index of NTB protection and tariff rates in the United States for samples of 290-300, 4-digit, SIC manufacturing industries, Godek (1985) found that high tariffs are associated with high NTB protection and that the share of tariffs in total protection declines as the overall level of protection increases. He found that protection in general is positively related to the share of sales going to final consumers and to the importance of unskilled labor in production. Nontariff trade restrictions are more likely to be found and to be higher for larger and geographically concentrated industries. Tariffs also increase with the share of sales to final consumers in an industry and decrease as the number of firms in an industry increases.

Stern and Deardorff (1985) took the analysis of tariff and NTBs a step further by using their Michigan model to assess the impact of tariffs and/or nontariff trade restrictions within a country when they are matched by similar forms of trade restrictions among foreign trading partners. The model includes information for 29 sectors, 22 tradeable goods and 7 nontradeables, for 34 countries. They focus on the implications of reductions in protectionist measures for Canada, Germany, Japan, Netherlands and the United States.

Stern and Deardorff find that the actual degree of protection afforded to an industry when foreign trade restrictions are taken into account can differ substantially from the impression one gains about protection when the foreign sector is ignored. For example, apparel and footwear become less protected for Canada. Metal products and electrical machinery are less protected for Germany. For Japan, tariffs were less protective for apparel, metal products, misc. manufactures, electrical machinery and transport equipment.

U.S. protection was found to be less for metal products, construction, petroleum, electrical machinery and chemicals. But, protection was greater for wood products, non-ferrous metals, textiles and leather products. NTBs reduced tariff protective effects for apparel, footwear, and textiles while increasing protective effects for rubber, metal and glass products and transport equipment. Obviously, special interests within an industry are interested in the net value of protectionist measures to them whether those measures are implemented at home or abroad. Future inquiries into the pattern of protection across industries should endeavor to define those net effects rather than simply focus on domestic tariffs and NTBs as measures of the degree to which domestic producers are protected from foreign competition.

Magee and Young (1987) have developed a model in which protection is endogenously determined that can be used to explain the pattern of changes in protection over time. They begin with the notion that Democrat administrations in the United States pursue macroeconomic policies that support high employment and high inflation domestically. Such policies would tend to cause a depreciation in the international value of the U.S. dollar. Those aggregate economic conditions tend to reduce import pressures and expand exports and are interpreted by Magee and Young to be conducive to freer trade conditions.

Alternatively they argue that Republican administrations pursue macroeconomic policies that are conducive to relatively low inflation rates, high unemployment rates and appreciation of the U.S. dollar, which in turn stimulates imports and discourages exports. The resulting increase in import pressure on the economy would tend to reenforce protectionist demands.

In combination, the patterns of behavior just described would result in the apparent paradox that protectionist demands and gains would be greatest during Republican administrations and least during Democratic administrations. While the empirical evidence offered in support for the hypotheses are less than conclusive, Leamer (1987), the ideas advanced by Magee and Young are both original and suggestive. At a minimum, the work by Magee and Young serves as a reminder that very little analytical and empirical work has been undertaken to explain changes in the pattern of protection over long periods of time. Even less research has been undertaken to explain and test the linkages between protectionism and monetary and fiscal policy choices among nations. More descriptive pieces that attempt to explain variations in protectionism in the United States over extended periods of time include Cassing, McKeown and Ochs (1986) and Ray (1987b, 1989b).

Cassing, McKeown and Ochs explain how the existence of regionally concentrated declining industries with sector specific capital in conjunction with the occurrence of business cycles could give rise to a cyclical pattern of protectionism. Protectionist pressures would be at their greatest and opposition weakest during economic downturns. Along somewhat different lines, Cassing and Hillman (1986) develop a model in which a declining industry falls below some threshold level of economic and political importance and protectionist support for the industry collapses. They suggest that something of the sort may have taken place with respect to the footwear industry in the United States.[8]

The discussion to this point has focused on broad patterns of protection across industries and on the extent to which tariffs and NTBs are coincident forms of protection or protect different industries. Based on several studies, we noted that NTBs appear to have been used in part to supplement tariff protection in those industries with the highest tariffs. On the other hand we noted that NTBs are associated with less concentrated industries, using capital-intensive methods of production which differs from the locus of concentration of tariff protection among industries. But, we have not discussed the characteristics or consequences of any particular form of NTB in any detail. While many of the cases involving protection for specific industries have involved administrative protection measures and therefore will be discussed in section III, there are a couple of cases of industry specific trade disputes worth mentioning in the current context.

The first case involves the use of voluntary export restraints, VERs, to limit exports of Japanese automobiles to the United States between 1981 and 1984. There have been a number of studies including Crandall (1985), Feenstra (1984, 1985) and Tarr and Morkre (1984) that have documented the impact of the agreement on employment and welfare in the United States and on the quality of automobile imports into the United States from Japan.

VERs represent a considerable threat to further efforts to liberalize world trade. As numerous authors have noted, VERs provide a means by which a government can provide protection for a domestic industry by bribing foreign governments and/or foreign producers with the tariff equivalent rents that are generated by the imposition of VERs. The only injured parties are the domestic consumers who pay more for protected goods than would otherwise be the case. There are no complaints to GATT. The government of the exporting country and the government of the importing country are co-conspirators in a scheme to reduce competition in the importing country, generate super-normal profits for producers in the exporting country and provide consumers in the importing country with less value per dollar spent.

Hillman and Urspring (1988) have developed an interesting model of lobbying efforts for protection in the presence of foreign subsidiaries. They find that as long as political campaigns and lobbying firms cannot capture tariff revenues there will always exist a preference among protectionist forces and politicians in the importing country for VERs rather than tariff protection.

In a recent study, Dinopoulos and Kreinin (1988) have added an important new twist to the story. They analyze the extent to which the VER agreement between the United States and Japan reduced overall competition in the U.S. automobile market and permitted automobile producers in Germany, Sweden and France to raise the prices of their cars in the U.S. market in order to capture extra profits.

First, Dinopoulos and Kreinin regressed unit values for automobiles on interest rates and unit labor costs for Germany, Sweden and France for the 1964-1980 period and forecast 1981-84 prices. They treated positive residuals as VER price mark-up effects. The weighted price mark-up on alternative foreign car imports into the United States ranged from $2,500 in 1981 to $6,900 in 1984.

As an alternative method of estimation of price mark-ups, they estimated hedonic price regressions and treated the difference between actual and estimated prices as equal to the VER mark-up. Following that approach, Dinopoulos and Kreinin found estimated mark-ups that ranged from $2,100 in 1981 to $6,600 for 1984.

Estimated welfare losses in the United States associated with the excess prices of Japanese automobiles ranged from $1.3 billion in 1981 to $2.4 billion in 1984 and are similar to findings in earlier studies. However, Dinopoulos and Kreinin also find substantial welfare losses for the United States associated with the excess prices on automobiles from third party suppliers in Europe. Those additional losses ranged from $0.9 billion in 1981 to $3.4 billion in 1984. Consequently, Dinopoulos and Kreinin estimated that it cost the U.S. economy $181,000 to save each autoworker job paying $35,000.

Kalt (1988) provides a separate industry study that illustrates the extent to which industry special interests can prevail at times in obtaining relief from competitive imports at the expense of the common good and even common sense. Lumber producers in the United States initiated a countervailing duty case against Canada that was rejected by the U.S. government in 1983.[9] Subsidized stumpage fees for removing trees from government controlled lands in Canada during the 1977-1984 period averaged only 14.6% of the market price for logging privileges in the United States. But, the fact that the subsidized trees were not used solely by lumber producers led the U.S. government to drop the case.

In 1983, the Commerce Department in the United States developed the "dominant use" principle to justify actions to protect the steel industry in the United States from imports from Brazil. Consequently, U.S. lumber producers launched another countervailing duty case in 1986 arguing that the dominant use for the trees that were cut at subsidized rates in Canada was to meet lumber production needs in Canada.

The International Trade Commission (ITC) recommended a countervailing duty of 15% to offset the estimated advantage in lumber production costs enjoyed by Canadian lumber producers because of the subsidized stumpage rate for clearing trees on government property in Canada. As Kalt argues, the evidence from the case suggests that logging companies captured virtually all of the quasi-rents associated with the stumpage pricing policy in Canada, which implies that the ITC misread the facts of the case.

In anticipation of the ITC ruling, the Reagan administration imposed a 35% tariff on wooden shakes and shingles imported into the United States from Canada in June, 1986. Canada responded by imposing tariffs on a variety of imports from the United States. In October, 1986 the United States imposed a 15% tariff on imports of soft lumber from Canada. Negotiations between the two countries resulted in a switch from the 15% tariff on imports into the United States to a 15% export tax on lumber from Canada on December 31, 1986. As Kalt indicates:

"What began as a large-country monopsony tariff directed against Canada (by the United States) has become a large-country monopoly tariff directed against the United States (by Canada)" (p. 342).

In its haste to satisfy the demands of the Coalition for Fair Lumber Imports in the United States, the U.S. government orchestrated a 15% tariff equivalent transfer to the Canadian government from American consumers of lumber based products. That negotiated giveaway by the U.S. government was implemented in order to offset a lumber production subsidy in Canada that did not exist. There are a number of other industry specific examples of trade restraints worth noting that will be addressed in the discussion of administrative protection in section III.

Legislative Votes and Administered Protection

While the discussion in section II touched on the politics of protection as well as the economics of protection, the focus was primarily on the economic circumstances that lead to protectionist pressures and on the attributes of pressure groups associated with effective coalitions. In a sense, those studies focus on the demand for protection. Pressure group characteristics help to determine whether a particular protectionist demand is likely to be effective or not.

Implicit in the notion of an effective demand for protection is a presumption of a potentially responsive source of supply. The primary focus of this section is on the supply of protection. Therefore, our attention will shift to the processes that lead to decisions to provide or deny protection. Those processes include voting on particular pieces of legislation sought by special interest groups and details about the administrative framework within which firms and industries can obtain relief from foreign competition.

Voting Behavior and Trade Legislation

In light of the recent partisan battle over trade policy that culminated in the Omnibus Trade and Competitiveness Act of 1988, it is useful to recall that in the immediate post-World War II period Democrats led the movement for trade liberalization while the Republicans tended to oppose movement in that direction. Baldwin (1976b) noted that on three trade bills between 1945 and 1950, 94% of Democrats voting in the House supported liberal trade bills while 92% of Republicans voting opposed the bills. When Republicans had a majority in both Houses of Congress in 1948 they pushed through a bill that established minimum tariffs below which domestic producers would be considered endangered by foreign competition (the so-called peril point tariff).

When the Democrats regained majorities in 1949, they quickly passed a trade bill that dropped the peril point tariff.

Baldwin recites the course of voting on trade bills throughout the 1950s and documents a drift toward protectionism that is generally missing from less thorough reviews of the post-war period. While protectionist pressures abated during the early 1960s, the Trade Expansion Act of 1962 was passed only in conjunction with extension of the petroleum import quotas initiated during the Eisenhower administration, the acceptance of a voluntary quota system for imports of cotton textiles, announcement of a six point program to assist the Northwest lumber industry, and increases in duties on imported carpet and glass under the escape clause provisions of the Trade Act of 1958. Labor and management joined forces in the steel industry in 1967 and gained a voluntary export restraint agreement with Japan and European steel producers in 1968.[10]

The Trade Act of 1974 was passed only after the trade adjustment assistance program was made easier to qualify for and more generous and the ability to retaliate against unfair trade practices was enhanced. The final vote in favor of the bill in the House was 272 to 140. But, party positions had reversed relative to the early post-war period. Republicans supported the basically liberal bill by a vote of 160 to 19 while Democrats opposed the bill by a vote of 112 to 121.

In an effort to quantify the influences behind voting behavior, Baldwin estimated regressions that related votes on the House bill in 1973 to a number of variables. He used probit analysis to estimate the contribution of various factors to opposition to the bill. Not surprisingly, Democrat House members opposed the bill. Opposition was also related to major union campaign contributions to congressional races in 1974 and to the relative importance of import sensitive industry employment in a House member's district. The presence of export industry interests in a congressional district did not influence votes significantly. Baldwin also found that the relative importance of specific industry interest groups within congressional districts such as oil and coal interests, which had benefitted from protection, contributed to the explanation of opposition to the trade bill.

To analyze Senate voting behavior, Baldwin considered votes in favor of an amendment that would have restricted the President from reducing tariffs for industries in which import shares were greater than one-third of domestic market sales in three of the previous five years. Members of the Senate who were Democrats and whose districts included a concentration of employment in import-sensitive industries supported the restrictive amendment.

A number of more recent studies have analyzed votes on more specific pieces of trade legislation. Coughlin (1985) analyzed House votes on the "Fair Practices in Automotive Products Act" better known as the domestic content legislation that passed in the House in October, 1982 by a vote of 215 to 188. The bill was ultimately vetoed by the President. the bill set minimum requirements for wholesale cost shares that had to originate in the United States. For example, a foreign auto producer selling 200,000 to 500,000 cars in the United States in 1983 would have been required to have a domestic content value of 25%. That requirement would have risen to 75% by 1985. Sellers of 500,000 or more units in The United States would have faced a domestic content requirement of 30% for 1983 and 90% for 1985. Non-compliance would have called for a 25% drop in sales permitted in the United States during the following year.

Coughlin tried to identify purely political factors influencing voting behavior, which he referred to as ideological factors. The measure that he used was the 1981 rating of House members by the Americans for Democratic Action, ADA. High ratings correspond to liberal votes on legislation, which in turn he expected to be associated with support for the bill.

Coughlin used a probit analysis and an n-chotomous probit to separately estimate whether or not House members supported the bill and the intensity of that support or opposition. He found that support for the bill was positively related to whether a voter was a Democrat, to the percentage steel industry employment in the district (24% of domestic use for steel in 1980 was for automobiles), to the percentage auto employment in the district, to the percentage unemployment in the district, to the percentage of labor contributions to campaign funds and to ADA ratings. Coughlin also estimated the marginal contribution of each variable to the overall explanatory power of the regressions. He found that the ideology measure did not add significantly to one's overall ability to explain voting behavior on the domestic content bill.

McArthur and Marks (1988) analyzed the same vote and began by replicating Coughlin's empirical results. They add the percentage of workers employed in export industries in the state, farm employment by state, a measure of consumer interest (auto registrations per 1,000 population by state) with the expectation that each would be positive. Only export employment was found to be significant. They also added the percentage of unionized labor in the workforce in the state and found it to be positive and significant as predicted. These changes caused the ideology measure and state unemployment rates to become insignificant.

Next, McArthur and Marks added the *National Journal* ratings of House members on economic issues in place of ADA ratings for 1981 and found that measure to be a significant factor in a positive vote (economic liberals supported the bill). Finally, they added a variable to reflect whether or not a House member had been re-elected in the election preceding the "Lame Duck" session in which the vote was taken. The result was positive and significant, which they interpreted as evidence that retiring and defeated members of the House voted their ideological preferences.

The interpretation of this last result can be questioned on two grounds. First, the authors may have inadvertently discovered a kind of lemons market phenomenon in politics. House members voting on the bill who did not get re-elected, perhaps because they were too dumb to satisfy constituent demands, still may not have figured out what their constituents wanted when they voted on the domestic content bill. In that sense, votes by non-returning members of the House may have little meaning in terms of ideological preferences.

Second, the vote itself was taken at a point in time when a veto that could not be over-ridden was anticipated. Yet, the only evidence of ideological voting on a bill that was doomed was the voting pattern of House members who would not be around for the next session of Congress. If it were in fact true that only lame duck representatives in lame duck sessions of Congress voting on dead-end bills vote their ideological preferences, then one could conclude that ideological preferences of legislators do not count for much.

Tosini and Tower (1987) analyzed House and Senate votes on the Textile Bill of 1985. The House version limited export growth to 1% per year for countries with more than a 1.25% share of the U.S. market in 1984 and to 6% otherwise. The Senate version reduced the number of exporters limited to the lower growth rate of sales in the United States from 12 countries to 3. The House passed the bill by a vote of 262 to 259 and the Senate passed the bill by a vote of 54 to 42. The final bill was vetoed by the President.

Using probit analysis to explain yes versus no votes in the Senate, Tosini and Tower found that a Senator's support was positively related to whether he/she was a Democrat, to the state percentage of employment in the textile industry, and to the unemployment rate in the state. Special interest campaign contributions, the percentage of the workforce in the state that is unionized, the percentage of employment in export industries and the length of time before the Senator faced re-election were all insignificant.

Voting in the House appeared to be responsive to more constituent characteristics. The likelihood that a House member voted for the bill was positively related to whether he/she was a Democrat, to special interest campaign contributions, to the percentage of employment in textiles in the state, and to the unemployment rate in the state. House support for the bill was negatively related to employment in export industries in the state.

Pugel and Walter (1985) analyzed trade legislation from a different vantage point. They generated samples of from 68 to 133 company responses to a survey instrument they sent to the 1980 Fortune 1000 group. Specifically, they asked about company attitudes toward the Burke-Hartke Bill, which was protectionist, the Trade Act of 1974, the Drafting of the 1979 Act, the 1979 Trade Act, the Results of the Tokyo Round and Future legislation. Their basic argument is that a company's position should be more protectionist the greater the threat from imports is in the domestic market, the less significant are the companies overseas sales interests, and the less diverse are the product markets served by the company.

Using ordered probit analysis they find general support for the proposition that companies are less protectionist in their attitudes toward trade legislation the more significant their export advantage, as measured by the average 1970-1979 R&D/Sales ratio, and more protectionist the higher tariff protection historically has been in the industry. Pugel and Walter find mixed evidence that production diversity is negatively related to protectionist positions, presumably reflecting less vulnerability to foreign competition in any one product line, and that historically high import penetration ratios in a companies major product markets are positively related to protectionist positions on trade legislation. Finally, using a composite index for each company that characterizes attitudes toward all of the legislation considered, they find protectionist company positions are positively related to tariff rates and import penetration in the companies markets and negatively related to the exporting potential of the company, and the diversity of the companies product offerings.

In a different context, Marvel and Ray (1987) tested the relationship between tariff protection across industries in the post-Kennedy Round period and the simultaneous presence of import and export interests in an industry. Their objective was to explore the relationship between intra-industry trade and protection. Using a sample of 314 4-digit SIC manufacturing industries they found that the simultaneous presence of export interests in import-sensitive industries was associated with reduced industry tariff protection.

Administered Protection

It is difficult to discuss technical rules associated with the administration of any program in an interesting way. Perhaps, it would be useful to remind ourselves of the cost to consumers and to the economy in
general of administered protection in the United States. Hufbauer,
Berliner and Elliott (1986) provided detailed estimates of the economic
consequences of special protection in 1984 for 31 industries. In general,
they noted the rapid spread of special protection measures across import competing industries. Special protection covered 8% of imports
valued at $12.4 billion in 1975 and 21% of imports worth $67.6 billion in
1984.

The annual costs borne by consumers in the United States from special
protection in specific industries included: Dairy Products $5.5 billion,
Automobiles $5.8 billion, Carbon Steel $6.8 billion, Phase III Textiles
and Apparel $27.0 billion. Estimates of jobs saved through special pro-
tection efforts include: Dairy 25,000, Automobiles 55,000 and Textiles
and Apparel 640,000.

To the extent that one is sympathetic about the plight of workers
who lose their jobs because of stiff foreign competition in the domestic
market, it is worthwhile reminding ourselves of the cost of protecting
jobs through protectionist measures. Hufbauer, Berliner and Elliott provide estimates of the annual costs to American consumers per job saved
in various industries. Those annual costs to consumers per job saved
include: Phase III Textiles and Apparel $42,000, Automobiles $105,000,
Motorcycles $150,000, Phase III Carbon Steel $750,000 and Specialty
Steel $1,000,000 (520 jobs saved). Clearly there is more involved in
these cases than saving jobs. A review of the empirical literature on
administered protection will clarify what some of the other considerations are in such cases.

Hillman (1989) provides a nice summary of administrative protection. For our purposes, we need only discuss three alternative routes to
special protection: escape clause, or safeguard cases, countervailing
duties cases and antidumping petitions. Escape clause cases are initiated by an appeal to the ITC for some form of relief from the competitive pressure of imports. Under the Trade expansion Act of 1962 an
industry had to provide evidence that imports were the primary cause
of economic distress. Beginning with the Trade Act of 1974, an industry
must simply provide evidence that it is faced with economic difficulties and that imports are a contributing factor. The ITC investigates the
industry claims of injury and if it agrees that relief in the form of a
tariff or some other form of protection or adjustment assistance is called

for, the ITC makes a recommendation to the President to assist the industry. The President's decision can be overridden by Congress if that decision is negative or if the President does not accept the remedy proposed by the ITC. But, there is no right to appeal a final decision in the federal courts.

Countervailing duty cases are filed with the International Trade Administration (ITA) of the Department of Commerce.[11] The ITA conducts an investigation to determine whether or not producers abroad benefit unfairly from direct or indirect government subsidies. The ITC also investigates whether or not evidence of injury or potential injury to the domestic industry exists. If evidence of unfair trade is established, the Commerce Department is required to impose a countervailing duty equal to the actual or implicit subsidy enjoyed by foreign producers. Alternatively, a negotiated settlement can be reached between the domestic petitioner and foreign exporter. Recall our discussion in section II of the lumber industry dispute, which resulted in a 15% export tax on lumber from Canada to the United States. In addition, the domestic petitioner can appeal a negative finding by the ITA in federal court.

Antidumping petitions are filed with both the ITA and the ITC. The ITC investigates whether material injury has been suffered or the potential for such injury exists for the domestic petitioner. The ITA investigates whether or not foreign producers are selling goods at "less than fair value" in the United States. If evidence is found to support the claims of material injury and sales at less than fair value, the Commerce Department is required by law to impose antidumping duties. As in the case of countervailing duties, there is room for a negotiated settlement among the parties involved in the dispute. As one might expect, the foreign producer is usually willing to raise the price of his/her product when the alternative is a tariff by the United States.

Before turning to a discussion of existing empirical work on administrative protection it is worth noting another form of administrative protection that will play a more prominent role in the future. Section 301 of the U.S. trade law authorizes the President to take steps to respond to violations of international trade agreements by trading partners if such violations are confirmed by an investigation of the U.S. Trade Representative. Prior to the Omnibus Trade and Competitiveness Act of 1988 complaints were initiated by a U.S. firm or industry with the office of the U.S. Trade Representative. The 1988 law requires the Trade Representative to take the initiative in identifying cases of violations, to report those cases to Congress, and to negotiate a settlement or implement a unilateral remedy within a specified period of time.

The first application of the revised law, referred to by some as "Super 301" led to reported trade violations by Brazil, India and Japan in June, 1989. The intent of the revised law is to force the U.S. Trade Representative to take an aggressive role in searching out and eliminating unfair bilateral trade practices. The first phase of the process, once offending parties have been named, is to attempt to negotiate a solution to the problems identified. Whether or not the revisions in section 301 will contribute to or obstruct future efforts to liberalize trade remains to be seen.

Finger, Hall and Nelson (1982) provide an interesting overview of how administrative protection works. They refer to escape clause cases as high track cases. By that they mean that escape clause cases are highly visible and politically charged. They refer to Less-Than-Fair-Value (LFV) cases, which they define to include both antidumping cases and countervailing duty cases, as low track cases. These are cases that turn on technical arguments and are subject to appeals procedures. Since LFV cases are initiated by firms and industries that want government assistance and to whom the government often must turn to for expert testimony on technical matters, Finger, Hall and Nelson argue that there is an inherent bias in favor of protectionist outcomes in LFV cases.

Finger, Hall and Nelson analyze the record of countervailing duty, escape clause and antidumping cases decided between 1975 and 1979. They analyze 183 of 208 possible cases of LFV pricing and 57 of 68 possible cases of LFV injury. Using logit analysis to test for the likelihood of an affirmative decision on LFV pricing cases they find that a positive decision is directly related to measures of comparative costs (assuming U.S. cost disadvantages are associated with industries characterized by a high capital/labor ratio, low wage, lack of scale economies in production) and to how precisely the affected product is defined i.e. how technical the case is to evaluate and negatively related to industry concentration. Concentration was hypothesized to be positively related to affirmative decisions.

A number of measures intended to capture political considerations including the 1976 share of U.S. exports to the country of the offending firm, a dummy variable reflecting bias against developing countries and industry size in the United States, all proved to be insignificant as expected.

LFV injury cases were also unaffected by foreign political variables, the comparative cost variables performed less well and technical precision was insignificant. Concentration was negatively signed. Positive decisions were associated with industry employment and a dummy variable for 1979, intended to test whether or not the Treasury Department made more favorable rulings in the hope of avoiding the eventual loss of authorization to handle these cases to the Commerce Department. Both of those variables were interpreted as domestic political factors.

One interpretation of the negative concentration effect could be that concentrated and politically powerful industries were taken care of with tariffs and other forms of protection and had no need to resort to LFV cases. Just as NTBs generally expanded the opportunities for less concentrated industries to gain relief from foreign competition, as discussed in section II, low track LFV cases may have added to their menu of protectionist options.

The average size of escape clause cases was much larger than the average size of LFV cases reflecting the greater importance of such cases hypothesized by Finger, Hall and Nelson. The average value of imports covered by a case was $331 million for escape clause cases and $106 million for LFV cases. If one excludes auto and steel antidumping cases that the authors argue did not belong in the LFV track and were not resolved there, the ratio of escape clause case size to LFV case size was 11 to 1, or, $331 million to $28 million.

The ITC made affirmative decisions in 25 of 40 cases. The President supported 8 of those 25 recommendations. Orderly Marketing Agreements (OMAs) were recommended by the President in 3 cases that accounted for 78% of the value of affirmative cases. Furthermore, Finger, Hall and Nelson found that the President decided in favor of 5 of 8 cases involving imports of $40 million or more but the President decided in favor of only 3 of 17 smaller cases. The presumption is that larger cases are politically harder to reject.

Baldwin (1985) analyzed the determinants of serious injury determinations by the ITC between 1974 and 1983. He finds that the percentage of commissioners voting who support a finding of injury in the 47 cases involved tended to increase the greater the loss in industry employment in the previous five years and the greater the decline in the net profits to sales ratio in the most recent two comparable periods (six months or two years). And, the simultaneous occurrence of both adverse industry changes improved the likelihood of favorable votes further. In effect, the ITC appeared to be finding evidence of industry injury in accordance with its mandate under the less stringent requirements of the Trade Act of 1974.

Baldwin also analyzed the determinants of Presidential decisions during the 1975-79 period. He found that a decision to provide industry relief by the President was not significantly related to the recommendation of the ITC. He does find that a favorable Presidential decision is more likely if unemployment rates are rising and the inflation rate is declining. And, Baldwin finds some support for the proposition that the President was more favorably disposed to provide relief for cases decided close to the time for Congressional elections and/or to the next Presidential campaign. These findings generally support the view by Finger, Hall and Nelson that escape clause cases are highly political matters, at least, once they reach the President's desk.

Takacs (1981) also analyzed escape clause cases. She reviewed petitions per year and the number of successful cases per year between 1949 and 1979. She expected to find that generally poor economic conditions (relatively low GNP, high unemployment, low capacity utilization rates) and deteriorating trade performance (a high import penetration ratio and overall negative trade balances) would increase both petitions and successful petitions per year. She also hypothesized that the shift to more lenient conditions for a favorable finding in the Trade Act of 1974 would increase petitions and successes and that petitions would be positively related to past successes. The empirical results are consistent with Takacs' hypotheses with respect to the rate of filing of petitions. The unemployment rate and the capacity utilization rate were not significant in explaining the number of successful cases per year. Baldwin's results regarding the political nature of affirmative decisions at the Presidential level may help to explain Takacs' greater success in explaining petitions per year compared to successful cases per year.

Hartigan, Perry and Kamma (1986) attempt to assess the impact of escape clause cases since the Trade Act of 1974 on the equity values of the petitioning firms. They perform event studies for 19 different industries using weekly stock values for firms in each industry beginning with stock prices 56 weeks before a petition was filed with the ITC and ending 4 weeks after a Presidential decision in cases with positive ITC findings. The overall impact on equity values of filing with the ITC for relief was significant in only 2 of the 19 cases. The results were positive in one case and negative in the other. Only one ITC decision, a negative one, had a significant stock effect and it was negative. One positive and one negative Presidential response with respect to import relief affected stock prices significantly out of 13 cases. But, both effects were negative.

Using the same basic techniques, Hartigan, Kamma and Perry (1989) analyzed the impact of antidumping proceedings on equity values for cases filed in accordance with the Trade Reform Act of 1979. The sample consists of 47 petitions involving 130 firms. They exclude steel industry cases that accounted for 50% of the filings since 1980.

In antidumping cases, the ITC has to make a preliminary finding of injury or the threat of injury. Then Commerce has to make a preliminary ruling on dumping. If that finding is affirmative, both Commerce and the ITC make final reports.

The event study analysis suggests that, absent a negative finding at some level, the equity effect of the filing process is positive. For threat of injury cases, the initial ITC ruling has a significant effect on equity values. For both injury and threat of injury cases the Commerce Department ruling has a significant effect on equity values. Analysis of variance tests indicate that residual equity gains associated with positive determinations in antidumping cases were significant only where injury was threatened and not when evidence of injury already existed. Presuming the antidumping mechanism delivers positive findings in the right cases, the event study evidence suggests that it is only useful if turned to in anticipation of losses and not after the damage is done.

Grossman (1986) provides evidence from an escape clause case that suggests that outcomes regarding administrative protection cases are not always right. Bethlehem Steel and the United Steel Workers of America filed with the ITC for escape clause relief in January, 1984. The petitioners contended injury from competitive imports during the period 1976 to 1983.

Grossman develops a reduced form equation for employment in steel mills and estimates industry employment. Simulations provide employment estimates when exogenous variables take on counterfactual values. Differences in employment figures associated with counterfactual versus actual values of exogenous variables are used to determine the employment effects of each of the exogenous variables. The reduced form employment equation is estimated using monthly data from January, 1973 to October, 1983. Exogenous factors include the relative prices of energy, iron-ore and the foreign price of steel each lagged 18 months, aggregate output and the wage rate in steel production lagged 5 months each and a trend variable. He develops counterfactual results for two sub-periods with the first beginning in January, 1976 and the second beginning in January, 1979.

Actual job losses in the steel industry between January, 1976 and August-October, 1983 equaled 144,367. Actual job losses between January, 1979 and August-October, 1983 equaled 169,167. Secular shifts alone, which are not explained in any detail, would have reduced employment by 208,734 in the first period and 109,600 during the second period. Job losses attributed to foreign competition during the two periods were 37,403 and 80,959, respectively. However, the major factor affecting import competition appears to have been the appreciation of the U.S. dollar. Grossman estimates that job losses for the period beginning in January, 1976 associated with currency appreciation accounted for 29,037 of the jobs lost through import competition. Currency appreciation accounted for 82,701 lost jobs compared to overall losses of 80,959 associated with import competition for the period beginning in January, 1979. In short, the case that job losses in the steel industry were attributable to foreign competition is not supported for the longer period used in the petition. And, the finding of substantial job losses due to foreign market effects for the shorter period would call for escape clause relief only if one believes that industries should be protected from the employment effects of currency movements through the use of such a program.

Conclusion

It is difficult, to say the least, to know best how to summarize a survey that has covered as much ground and taken up as much space as this one. For the sake of brevity, let me simply summarize what we know about the structure and determinants of protection based on our review of the empirical literature. That summary should suggest areas for further research and policy issues that may become more important in the next decade.

Tariff protection tends to be associated with industries in which a country has a long standing comparative disadvantage in trade. For a country like the United States, such industries tend to use low-skill, labor-intensive methods of production and to produce products such as consumer goods, textiles and processed agricultural products. Evidence that market characteristics affect the distribution of tariff protection is mixed. There is some evidence to suggest that concentration of production among a few firms and geographical concentration of production are characteristic of industries with relatively high tariffs.

Tariff protection is more likely to be associated with declining industries than with expanding industries in industrialized countries and tariff concessions, at least by the United States, over the last 60 years have not systematically altered the basic structure of tariffs across industries. Tariff cuts associated with GATT Rounds have been biased away from commodity groups in which the developing countries have the greatest potential for exports to the industrialized countries. Efforts to provide developing countries with compensatory, duty-free access to industrialized country markets with programs like the GSP have not been successful.

There is no evidence to indicate that tariffs have had a strong impact on a given country's trade balance. There is evidence to suggest that the protective effects of domestic tariffs can be misleading because of the compounding effects of protection abroad as well as NTBs at home.

NTBs have been used to supplement tariff protection in the most highly protected sectors of industrialized areas, including the United States, Canada, the European Community and Japan. Commodity groups that enjoy both tariff and NTB protection in the industrialized areas include consumer goods, textiles and processed agricultural products. NTBs have also been used to provide new trade restrictions in traditionally less protected areas as well, including standardized products using capital intensive methods of production.

There is some evidence to suggest that NTBs do influence bilateral trade flows but not overall trade balances for the industrialized countries. In addition, NTBs in one country may induce NTB protection in trading partner markets. There is evidence that domestic policies regarding depreciation allowances, legal treatment of vertical restraints, cartels and other market phenomena can influence bilateral trade flows.

To the extent that protectionism is an increasing threat, the threat comes from the proliferation in the use of NTBs. NTBs tend to be associated with less concentrated industries suggesting that industries that traditionally were unable to form effective lobbies to gain tariff protection can gain trade restraints in the form of NTBs. That implies that innovations in the use of NTBs, other things equal, have tipped the balance between protectionist and trade liberalizing forces in the direction of protection.

Voting on particular pieces of protectionist legislation tends to reflect the economic interests of a legislator's constituents. House member votes in Congress generally support the positions of industries that are important within representatives' districts. Special interest campaign contributions, union strength within a state and state unemployment all promote protectionist voting in the House and to a lesser extent in the Senate. Export interests within a state tend to be less effective in influencing votes on particular bills. Although, there is some evidence that the presence of export interests in import-sensitive industries has moderated the amount of tariff protection in an industry.

Purely political or ideological influences on legislative votes with respect to the regulation of trade have not been documented and the prognosis for success in isolating such influences is not good. That does not imply that there are not other issues for which more purely political concerns are important in explaining voting behavior.

Administrative protection, as structured in the United States, seems to be biased toward yielding protectionist outcomes in countervailing duty and antidumping cases. Escape clause cases appear to be determined at the ITC level on the basis of evidence of economic hardship within an industry. Presidential decisions on escape clause cases appear to reflect a greater willingness to provide trade relief in larger cases than in smaller ones and are not bound by ITC recommendations.

The use of the administrative track to obtain protection is associated with less concentrated industries. In effect, that evidence may simply reflect the fact that all forms of NTBs including administrative protection have expanded the menu of opportunities available to industries seeking relief from foreign competition and reduced the organizational costs required for success. If so, holding economic and political conditions constant, protectionism is a greater threat to the world economy now than in the past.

Efforts to reform trade policy such as calls for retariffication, converting NTBs into tariffs, may prove terribly naive. The current mix of tariffs, quotas, VERs, countervailing duty, antidumping, escape clause and other NTB arrangements is not the product of some random process. Industries that have gained trade relief through administrative protection are aware of their own inability to sustain support for tariff relief and would presumably fight any attempt to convert what they have for what they cannot hold onto in the future.

Specific studies of trade relief in the form of NTBs including automobiles, lumber and steel tend to reflect a triumph of industry interests over consumer interests with costs to consumers that are many times the value of the job losses prevented in the given industry. All three cases challenge government findings regarding relevant economic facts and, therefore, reflect poorly on the quality of analyses that have served as justification for government intervention in those markets.

Furthermore, Grossman's study of the steel industry highlights the important role that macroeconomic policies play in determining the strength and effectiveness of protectionist demands. He provides convincing evidence that exchange rate movements played the most significant role among identifiable factors in explaining employment decreases in the U.S. steel industry between 1979 and 1983.

Let me conclude by listing a few key areas of research on the political economy of protection for which empirical research is lacking or inadequate. 1) The relationship between protection in one country and the adoption of protection by trading partners, 2) The impact of a country's

protectionist measures on domestic resource allocation given protectionist measures abroad and exchange rate policies at home and abroad, 3)
Third party responses to bilateral trade agreements, as suggested by
the auto case in the United States, 4) The distribution of firms and industries among the various forms of trade restraints available, 5) The
evolution of innovations in protection including who innovates and
why, and 6) The linkages between macroeconomic policies and industry
specific protection.

Notes

1.   Although, as Robert Baldwin (1985) and others have pointed out,
there is a great deal of ambiguity about the interpretation of empirical
results based on the significance of proxy variables. Too often empirical
tests are not strong enough to distinguish among competing hypotheses.

2.   Leontief's Paradox refers to the empirical finding that the
United States tends to import capital-intensive goods and export labor-
intensive goods despite the abundance of physical capital in the
United States relative to the rest of the world in the early post-World
War II period.

3.   Baack and Ray provide evidence that tariffs were
systematically applied in industries facing inelastic consumer demand,
presumably to enhance federal government revenues. They also found
that every major manufacturing sector that emerged in the United
States by World War I had been systematically protected throughout
the late nineteenth century.

4.   The conservative social welfare bias explanation of government
behavior in the trade area is generally attributed to Max W. Corden
(1974).

5.   The absence of a private insurance market for business failures is
explained by the significance of the moral hazard problem. Firms in-
sured against failure have less incentive to avoid failure than would
otherwise be the case. The argument for public provision of business
insurance against failure would require a rationale of its own. For
example, perhaps the lack of such insurance leads private en-
trepreneurs to avoid risks that would be profitable from the perspective of society as a whole.

6.   The free-rider problem associated with tariff protection arises
from the fact that domestic firms benefit from tariff protection
whether or not they contribute to the coalition efforts that secured protection. There is an incentive to free-ride on the efforts of others. NTBs

can be used to provide additional rents and therefore incentive for individual firms to participate in lobbying efforts.

7. Baldwin also provides fresh evidence regarding the determinants of administered protection that we will defer discussion of to section III.

8. The Bush administration proposals for the environment in June, 1989 may reflect the beginning of a collapse in political support for the coal industry in the United States.

9. A detailed discussion of how countervailing duty cases are adjudicated is presented in section III and is not essential for an understanding of the developments in the lumber case.

10. The agreement was for a three year period and was renewed in 1971 for a second three year period.

11. Prior to 1980, cases were filed with the Treasury Department. Congress shifted administrative responsibility for investigating such cases to Commerce presuming that the Commerce Department would be more responsive to the "will" of Congress then the Treasury Department.

References

Anderson, Kym. 1980. The Political Market for Government Assistance to Australian Manufacturing Industries. *Economic Record*. 56:132-44.

Audretsch, David B. and H. Yamawaki. 1988a. R and D Rivalry, Industrial Policy and U.S.-Japanese Trade. *Review of Economics and Statistics*. 70:438-447.

———. 1988b. Import Share Under International Oligopoly with Differentiated Products: Japanese Imports in U.S. Manufacturing. *Review of Economics and Statistics*. 70:569-576.

Baack, Bennett D. and E.J. Ray. 1983. The Political Economy of Tariff Policy: A Case Study of the United States. *Explorations in Economics History*. 20:73-93.

Balassa, Bela. 1967. The Impact of the Industrial Countries' Tariff Structure on their Imports of Manufactures from Less Developed Areas. *Economica*. 34:372-83.

Baldwin, Robert E. 1971. Determinants of the Commodity Structure of U.S. Trade. *American Economic Review*. 61:126-46.

———. 1976a. Trade and Employment Effects in the United States of Multilateral Tariff Reductions. *American Economic Review Papers and Proceedings*. 66:142-48.

———. 1976b. The Political Economy of Postwar U.S. Trade Policy. Bulletin, New York University Graduate School of Business, No. 4.

————. 1982. The Political Economy of Protectionism. *Import Competition and Response.* Jagdish Bhagwati, ed. p. 263-286. Chicago: University of Chicago Press (for N.B.E.R.) (and comments by Stephen Magee and Stanislaw Wellisz p. 287-292).

————. 1984a. Rent Seeking and Trade Policy: An Industry Approach. *Weltwirtschaftliches Archiv.* 120:662-677.

————. 1984b. Trade Policies in Developed Countries. In *Handbook of International Economics.* Ronald Jones and Peter Kenen, eds., p.571-619. Amsterdam North-Holland.

————. 1984c. The Changing Nature of U.S. Trade Policy Since World War II. *The Structure and Evolution of Recent U.S. Trade Policy.* Robert E. Baldwin and Anne O. Krueger, eds., p. 5-27. University of Chicago Press (for N.B.E.R.).

————. 1985. *The Political Economy of U.S. Import Policy.* Cambridge, Mass: MIT Press.

Becker, Gary. 1976. Comment. *Journal of Law and Economics.* 19:245-48.

————. 1983. A Theory of Competition Among Pressure Groups for Political Influence. *Quarterly Journal of Economics.* 98:317-400.

Brock, William and S.G. Magee. 1978. The Economics of Special Interest Politics: The Case of the Tariff. *American Economic Review.* 68:246-50.

Cassing, James H. and A.L. Hillman. 1986. Shifting Comparative Advantage and Senescent Industry Collapse. *American Economic Review.* 76:516-523.

Cassing, James H., T.J. McKeown and J. Ochs. 1986. The Political Economy of the Tariff Cycle. *American Political Science Review.* 80:843-862.

Caves, Richard E. 1976. Economic Models of Political Choice: Canada's Tariff Structure. *Canadian Journal of Economics.* 9:278-300.

Cheh, John H. 1976. A Note on Tariffs, Nontariff Barriers, and Labor Protection in United States Manufacturing Industries. *Journal of Political Economy.* 84:389-94.

Chow, Peter C.Y. and M. Kellman. 1988. Anti-LDC Bias in the U.S. Tariff Structure: A Test of Source Versus Product Characteristics. *Review of Economics and Statistics.* 70:648-53.

Clark, Don P. 1987. *Regulation of International Trade: The United States' Caribbean Basin Economic Recovery Act.* University of Tennessee, July.

Corden, Max W. 1974. *Trade Policy and Economic Welfare.* Oxford: Clarendon Press.

Coughlin, Cletus. 1985. Domestic Content Legislation: House Voting and the Economic Theory of Regulation. *Economic Inquiry.* 23:437-48.

Crandall, Robert. 1985. *Assessing the Impact of the Automobile Voluntary Export Restraints upon U.S. Automobile Prices.* The Brookings Institution, October.

Deardorff, Alan V. and R.M. Stern. 1979. American Labor's Stake in International Trade. In *Tariffs, Quotas and Trade: The Politics of Protectionism.* San Francisco: Institute for Contemporary Studies, p. 125-48.

———. 1985. The Structure of Tariff Protection: Effects of Foreign Tariffs and Existing NTBs. *Review of Economics and Statistics.* 67:539-548.

Dinopoulos, Elias and M. Kreinin. 1988. Effects of the U.S.-Japan Auto VER on European Prices and on U.S. Welfare. *Review of Economics and Statistics.* 70:484-91.

Feenstra, Rob. 1984. Voluntary Export Restraint in U.S. Autos, 1980-81: Quality, Employment and Welfare Effects. *The Structure and Evolution of Recent U.S. Trade Policy.* R. Baldwin and A. Krueger, eds. Chicago: University of Chicago Press, N.B.E.R.

———. 1985. Automobile Prices and Protection: The U.S.-Japan Trade Restraint. *Journal of Policy Modeling.* 7:49-68.

Finger, Hall and Nelson. 1982. The Political Economy of Administered Protection. *American Economic Review.* 72:452-66.

Godek, Paul. 1986. The Political Optimal Tariff: Levels of Trade Restrictions Across Developed Countries. *Economic Inquiry.* 24:387-93.

———. 1985. Industry Structure and Redistribution Through Trade Restrictions. *Journal of Law and Economics.* 28:687-703.

Goldstein, Judith L. and S.D. Krasner. 1984. Unfair Trade Practices: The Case for a Differential Response. *American Economic Review Papers and Proceedings.* 74:282-87.

Grossman, Gene. 1982. Import Competition for Developed and Developing Countries. *Review of Economics and Statistics.* 64:271-81.

Grossman, Gene M. 1986. Imports as a Cause of Injury: The Case of the U.S. Steel Industry. *Journal of International Economics.* 20:201-223.

Hartigan, James C. and E. Tower. 1982. Trade Policy and the American Income Distribution. *Review of Economics and Statistics.* 64:261-270.

Hartigan, James C., P.R. Perry and S. Kamma. 1986. The Value of Administered Protection: A Capital Market Approach. *Review of Economics and Statistics.* 68:610-617.

Hartigan, James C., S. Kamma and P.R. Perry. 1989. The Injury Determination category and the Value of Relief from Dumping. *Review of Economics and Statistics.* 71:183-86.

Helleiner, G. K. 1977. The Political Economy of Canada's Tariff Structure: An Alternative Model. *Canadian Journal of Economics.* 10:318-326.

Hillman, Arye L. 1982. Declining Industries and Political-Support Protectionist Motives. *American Economic Review.* 72:1180-1187.

———. 1989. *The Political Economy of Protection.* Chur: Harwood Academic Publishers.

Hillman, Arye L., and H.W. Ursprung. 1988. Domestic Politics, Foreign Interests, and International Trade Policy. *American Economic Review.* 78:729-45.

Hufbauer, Gary C. 1970. The Impact of National Characteristics and Technology on the Commodity Composition of Trade in Manufactured Goods. *The Technology Factor in International Trade.* Raymond Vernon, ed. New York: Columbia University Press (for the National Bureau of Economic Research).

Hufbauer, Gary, D. Berliner, and K.A. Elliott. 1986. *Trade Protection in the United States: Thirty-one Case Studies.* Institute for International Economics, Washington, D.C.

Kalt, Joseph P. 1988. The Political Economy of Protectionism: Tariffs and Retaliation in the Timber Industry. *Trade Policy Issues and Empirical Analysis.* Robert E. Baldwin, ed. University of Chicago Press. NBER:339-364.

Lavergne, Real. 1983. *The Political Economy of U.S. Tariffs: An Empirical Analysis.* New York: Academic Press.

Leamer, Edward E, 1987. Comment on 'Endogenous Protection in the United States, 1900-1984.' *U.S. Trade Policies in a Changing World.* Robert M. Stern, ed., p.196-200. Cambridge Mass: MIT Press.

Magee, Stephen P. and L. Young. 1987. Endogenous Protection in the United States, 1900-1984. *Trade Policy in the 1980's.* Robert M. Stern, ed., p.148-195. Cambridge, Mass.: MIT Press.

Marvel, Howard P. 1980. Foreign Trade and Domestic Competition. *Economic Inquiry.* 18:103-22.

Marvel, Howard P. and E.J. Ray. 1983. The Kennedy Round: Evidence on the Regulation of International Trade in the United States. *American Economic Review.* 73:190-97.

———. 1987. Intraindustry Trade: Sources and Effects on Protection. *Journal of Political Economy.* 95:1278-91.

McArthur, John and S.V. Marks. 1988. Constituent Interest vs Legislator Ideology: The Role of Political Opportunity Cost. *Economic Inquiry.* 26:461-470.

Nelson, Douglas R. 1981. *The Political Structure of the New Protectionism.* World Bank, staff working paper no. 471.

Olson, Jr., Mancur. 1968. The Logic of Collective Actions: *Public Goods and the Theory of Groups.* New York: Schocken.

Peltzman, Sam. 1976. Toward a More General Theory of Regulation. *Journal of Law and Economics.* 19:211-40.

Pincus, J.J. 1975. Pressure Groups and the Pattern of Tariffs. *Journal of Political Economy.* 83:757-78.

Pugel, Thomas A. and I. Walter. 1985. U.S. Corporate Interests and the Political Economy of Trade Policy. *Review of Economics and Statistics.* 67:465-73.

Ray, Edward John. 1974. The Optimum Commodity Tariff and Tariff Rates in Developed and Less Developed Countries. *Review and Economics and Statistics.* 56:369-77.

———. 1981a. The Determinants of Tariff and Nontariff Trade Restrictions in the United States. *Journal of Political Economy.* 89:105-21.

———. 1981b. Tariff and Nontariff Barriers to Trade in the United States and Abroad. *Review of Economics and Statistics.* 63:161-68.

———. 1987a. The Impact of Special Interests on Preferential Tariff Concessions by the United States. *Review of Economics and Statistics.* 69:187-93.

———. 1987b. Protectionism: The Fall of Tariffs and the Rise in NTBs. *Northwestern Journal of International Law and Business.* 8:285-325.

———. 1989a. The Impact of Rent Seeking Activity on U.S. Preferential Trade and World Debt. *Weltwirtschaftliches Archiv.* 125:619-38.

———. 1989b. U.S. Protectionism and the World Debt Crisis. Quorum Books.

Ray, Edward John and H.P. Marvel. 1984, The Pattern of Protection in the Industrialized World. *Review of Economics and Statistics.* 66:456-58.

Saunders, Ronald S. 1980. The Political Economy of Effective Protection in Canada's Manufacturing Sector. *Canadian Journal of Economics.* 13:340-348.

Staiger, Robert W., A.V. Deardorff and R. Stern. 1988. The Effects of Protection on the Factor Content of Japanese and American Foreign Trade. *Review of Economics and Statistics.* 70:475-83.

Stern, Robert M. 1963. The U.S. Tariff and the Efficiency of the U.S. Economy. American Economic Review Papers and Proceedings. 54:459-70.

Stigler, George J. 1971. The Economic Theory of Regulation. *Bell Journal of Economics and Management Science.* 2:3-21.

Stone, Joe A. 1978. A Comment on Tariffs, Nontariff Barriers, and Labor Protection in United States Manufacturing Industries. *Journal of Political Economy.* 86:959-62.

Summary, Rebecca. 1989. A Political-Economic Model of U.S. Bilateral Trade. *Review of Economics and Statistics.* 71:179-82.

Takacs, Wendy. 1981. Pressures for Protectionism: An Empirical Analysis. *Economic Inquiry.* 19:687-93.

Tarr, D.G. and M.E. Morkre. 1984. *Aggregate Costs to the United States of Tariffs and Quotas on Imports.* Washington D.C.: Federal Trade Commission.

Taussig, Frank. 1931. *The Tariff History of the United States.* New York: Putnam.

Tosini, Suzanne and E. Tower. 1987. The Textile Bill of 1985: The Determinants of Congressional Voting Patterns. *Public Choice.* 54:19-25.

Discussion

Vernon Roningen

First, I would like to see similar summaries of empirical work on the determination of trade policy in other countries, particularly the European Community and Japan. I realize that some of the studies surveyed do deal with other countries, including Japan. However I think that a comparative survey might be especially useful to scholars and policy makers everywhere as they try to negotiate away or otherwise deal with each others' trade barriers. It certainly would be interesting to see studies done with similar methodologies for different countries. As the world shrinks, interest groups who are seeking economic rents via the political process in various countries are getting to know each other; even they might be interested in reading such a survey. Those in government who want to formulate and operate trade policy in the future need to know not only the economics and numbers associated with policies, but also the politics needed to get trade policies in place (or stop them in their tracks, if that is the task at hand).

In the United States that means understanding, and dealing with the Congress, lobbyists, and interest groups. In trade negotiations, you need similar knowledge about the countries you negotiate with. For example, I simply do not know if a similar type of economic literature on the political and economic determinants of protection in Europe exists. If it does, it should be surveyed and if not, it should certainly be created.

Second, I would like to ask and comment upon the obvious question "What is the practical application of this type of research?" Empirical studies on esoteric topics such as the "the impact of the Export Enhancement Program" do have some practical uses. Sometimes politicians want to know what tax dollars are doing or if laws are being faithfully executed (oversight). Other times, administrations want to know if programs are working or if not, how to fix them. Numbers, charts, and measures of economic rents are key weapons of economic war in the halls of Congress. I think it is more difficult to excite politicians and bureaucrats with statements such as "there is some evidence to suggest that concentration of production among a few firms and geographical concentration of production are characteristic of industries with relatively high tariffs." Now if such a statement can be translated to "Who do I call to get support for my bill?" or "Is my state getting its fair share of Federal benefits?," then we are getting somewhere.

As one answer to my own query, I suggest the following: Types of studies cited in the paper are the most valuable for the viewpoints on the policy process that they generate. If anything is going to help smooth the way toward a more open world economy and liberalized trade, I think it will be the ability of all sides to look at each others policies and economic and political systems that generate those policies. For example, I found the comment on the Stern and Deardorff work quite interesting where they say that if one nets out foreign trade restrictions, a country's "actual degree of protection . . . can differ substantially from the impression one gains about protection when the foreign sector is ignored." The ability to view the protection problem from this perspective can truly be of help to anyone pursuing trade negotiations.

Third, what should be done in this area in the future? I generally agree with the items in Professor Ray's list at the end of his paper. I think it is extremely important to empirically look at protection across countries in the same industry if we are to pursue the "disarmament" approach to trade negotiations. We have had some success in agriculture in measuring support on a roughly comparable basis across countries. On balance, I think this work has finally made the multilateral reduction of support to agriculture a possibility. I might add that ignoring the political environment of our negotiating partners has made that possibility remote. Simply put, if you can't measure it, you can't negotiate it. If you can measure it but don't understand the political realities behind the protection, you won't be able to negotiate anything.

If there were an institution which had the mandate to annually calculate the rents received by each industry in the U.S. economy from protection, we would have a better more rational trade policy. If such rents were calculated across industries in the major trading countries, we might find it easier to negotiate reductions internationally. And if such rent calculations were available, scholars doing empirical research on the political economy of trade would have a much better data base to work with.

The other item in the list I deem critical is the linkage between macroeconomic policies, including the flexible exchange rate, and industry protection. To put it simply, if business cycles affect some sectors of the economy more than others, you can be sure that political systems, including ours, will try to protect those sectors most vulnerable. It would be interesting to see some research in this area distinguishing between protection from the effect of some other policy (e.g., macroeconomic) and protection established just to capture economic rents.

I would add something to the list of items for further research that is mentioned in the survey. That is the idea that a very broad view needs to be taken regarding protection across countries. The cultures and history in different market economies may give rise to different rules which define "protection." Such situations have to be accounted for in our economic and political calculus if we going to pursue open, rather than cartelized, world markets in the future.

Viewed broadly, protection in its various forms, is a long-standing form of industrial policy which needs to be made more transparent. The world is shrinking, markets are integrating, economic change is accelerating, policies and their effects are becoming more transparent, and trade policy will evolve into world sector or industrial policy. This has become true for agriculture as well as other sectors of the economy where governments intervene heavily.

We all need the perspective this paper gives us so we can deal with all of the views that make trade and agricultural/industrial policy in the United States and our trading partners. But having worked in a Congressional office, I must also emphasize that personalities and situations matter very much in the outcome of any legislation with an economic impact. If you read the paper in Washington at the final passing of an annual appropriations bill you probably will come across editorials commenting on the "pork barrel" politics involved in the bill. However if you read a state or local newspaper on the same bill, you will see comments about "bringing home the bacon" for the home state. Given our constitution and governmental system, that is the way it is. The market of Congressional votes is a relatively thin one, especially at the committee level, and economists have never had spectacular success in explaining price behavior in thin markets. Personalities, particular situations and tradeoffs, and partisan and regional conflicts in the Congress, also matter a lot in the final outcome of economic and trade legislation.

8

The Political Economy of Agricultural Policy and Trade

Harry de Gorter and Yacov Tsur

Introduction

Government policy interventions in agriculture vary over time and across countries and commodities with a myriad of instrumentalities being employed. Several stylized facts on the patterns of government intervention in world agriculture have emerged including the fact that subsidies to farmers increase in countries with higher levels of GNP or industrialization while developing countries tend to tax farmers.[1] Agricultural protection has increased in industrial countries in the past several decades while that for manufacturing has declined. The taxation of agriculture in many low income countries has occurred in the face of a large rural population while rural votes in industrial countries are relatively fewer.

Alternative explanations for this pattern of intervention have been advanced including social welfare maximization,[2] "class" theories of special interests using the state for their own benefit,[3] and alternative theories of interest group behavior.[4] Olson (1985, 1986) for example argues that the rural sector in developing countries is exploited because the large and dispersed members of this sector can neither organize themselves adequately to pressure the government to act on their behalf. In industrial countries, the urban sector is large and dispersed and hence exploited for the benefit of an organized and smaller rural sector.

Becker's (1983) theory of pressure groups argues that relatively small groups are successful in obtaining subsidies in agriculture. Becker assumes passive behavior by politicians who are assumed merely to transmit the pressure of active groups. Competition between pressure groups is the primary determinant of political influence in Becker's framework. Gardner adopts this model by focusing on the cost of generating pressure by farmers in U.S. agriculture.

This paper adopts the model by Downs (1957) and Breton (1974) of rational, self-interested behavior by politicians, developed formally by Frey and Lau (1968), and Peltzman (1976) and applied to agriculture in de Gorter. Politicians or political parties compete for support defined in terms of votes, popularity ratings in polls and other measures that reflect the intensity of voter preferences for the government. In order to achieve and maintain power, governments maximize a "political support" function in choosing the level of intervention in agricultural markets. Politicians formulate policies in order to be in power and not vice-versa. The logic of political support from individuals is integrated with a logic of decisions made by politicians whereby price policy decisions depend on political support and vice-versa. For simplicity, we assume two interest groups, composed of agricultural producers (rural sector) and consumers/taxpayers (urban sector). These interest groups are assumed to be passive and merely transmit their preferences for or support of the government. The political support functions are specified to depend on *relative income* between groups and *redistributed income* within groups. Agricultural policies are viewed as equity-motivated to deal primarily with poverty and "basic needs" in developing countries (Schuh 1978) and large adjustment costs for farmers in industrial countries.

The model has politicians exploiting the differences in the marginal utility of income resulting from differing endowments such that redistribution allows governments to maximize political support. In industrial countries, the adjustment costs for the agricultural sector are high because of a low income elasticity of demand for food as income rises, resulting in the so-called "farm problem." In developing countries, on the other hand, agriculture is burdened with the consequences of equity-motivated policy designed to satisfy urban consumers demand for "cheap food" because it is a major share of their budget expenditures.

The results of the model developed in this paper show that per capita transfers depend critically on relative endowment (pre-policy) incomes and numbers in each group. Conditions are derived to determine when farmers are taxed or subsidized, and when over-shooting occurs, i.e., when government transfers result in a switch in the relative incomes between groups. Because political support is also a function of the change in income within groups due to redistribution, the tendency for governments to reduce the disparity in relative incomes is partially mitigated. It is also determined that the marginal deadweight loss of redistribution affects the equilibrium transfer. Hence, commodity sectors that have more elastic demand and a more inelastic supply are expected to have higher subsidies or lower taxes while farmers in import competing sectors will receive greater subsidies (or lower taxes) than those in an export sector. A change in the relative numbers in each group also affects the equilibrium transfer but in two offsetting ways. For example, a decrease in the relative number of farmers increases the per capita transfer because there are fewer farmers to subsidize and relatively more urban members to tax; but a smaller rural population becomes relatively less important to governments in terms of political support so that the transfer would decrease.

The pre-policy income gap between the urban and rural sector is hypothesized to be higher in industrial countries and hence the model partially explains why industrial countries subsidize farmers while developing countries do the opposite. Endowment income differentials between the urban and rural sectors is one explanation for why West Germany (with many farmers) prefers higher cereal prices than other countries in the European Community while the United Kingdom opposes price increases (even with fewer and larger farms). The model also explains via pre-policy relative endowment incomes why Canada and Argentina as two urbanized countries with similar agro-climatic conditions have diametrically opposite agricultural price policies and why non-food export sectors with fewer and richer farms in developing countries are taxed significantly more than farmers in the more populous staple food production sectors. The model is in contrast to the interest group models of Olson (1985, 1986) and Becker who argue that world agricultural price bias necessarily favor farmers in those sectors and countries that have relatively fewer farms.

Patterns of Intervention in World Agriculture

The taxation of agriculture in developing countries and subsidization of farmers in industrial countries has become one of the more predominant patterns of government involvement in agriculture observed by economists. The evidence suggests that subsidies increase with the level of GNP or industrialization. Anderson and Hayami (1986) observe that countries in South-East Asia and Europe shift from taxing to subsidizing agriculture in the course of economic development and industrialization. The gap between manufacturing and agricultural productivity increases in the process of industrialization as the urban population gets larger and richer. In the cases where agricultural productivity is also high in industrial countries, demand is inevitably inelastic and contributes to the adjustment problems in agriculture. Protection for the manufacturing sector has decreased in industrial countries and increased in developing countries in the past four decades (Anderson and Hayami 1986). In poor agrarian economies, on the other hand, manufacturing productivity and wages are lower compared to industrial countries. In addition, the agricultural sector is relatively larger in terms of both GNP and population. Hence, it becomes easier politically to tax the larger and richer urban sectors in industrial countries where food is a low percentage of total expenditures and subsidize farmers who are fewer in number and coping with structural adjustment problems.

Anderson and Tyers (1988) determine a correlation between agricultural protection and per capita national income and conclude that society has an income elastic demand for assisting farmers. Honma and Hayami (1986) find a statistical correlation between agricultural protection and both its comparative advantage and international terms of trade relative to the manufacturing sector. Anderson and Tyers note several exceptions to these patterns such as food-exporting rich countries like Australasia and North America. Our model gives some possible explanations for this later.

Krueger, Schiff and Valdes (1988) determine the direct (sectoral) and indirect (macro) policy impacts on incentives in agriculture in eighteen developing countries. Farmers are taxed through import-substitution policies in the industrial sector, over-valued exchange rates via exchange-control regimes and import licensing, and suppressed farm prices via government procurement policies (especially marketing boards), export taxes or quotas. Some of these taxes have been offset by subsidies to inputs, irrigation and the like. Direct protection was found to be negative for exportables (typically non-food crops) but positive (with exceptions) for importables (often food staples). Total protection averaged -7 percent for importables and -35 to -40 percent for exportables. Similar conclusions were reached earlier in a study by Kerr (1985).

A Theoretical Model

We now develop a model that explains government behavior in setting price policy. Underlying the model is the notion of a member's "intensity of political support," interpreted as the probability that a maximize political support, we derive results on the government's rules of behavior. We shall compare these results with some evidence compiled on the patterns of government intervention in agriculture.

Consider an economy that consists of two homogeneous groups: urban and rural of sizes n_u and n_r, respectively. Let T denote the total income transfer from the urban to the rural sector (a negative T denotes an income transfer from the rural to urban sector) and let $t = T/n_u$ be the per capita urban tax. The per capita subsidy to the rural sector is thus $t_r = T/n_r = tR$, where $R = n_u/n_r$, the population ratio between the urban and rural sectors. Let $Y_0(t)$ and $\Pi_0(t)$ denote the urban and rural per capita endowment income net of transfer payments t; thus

$$Y(t) = Y_0(t) - t \tag{1}$$

$$\Pi(t) = \Pi_0(t_r) + t_r \tag{2}$$

are the total realized per capita incomes in the urban and rural sectors.

The transfer T distorts individual choices. The result is the (per capita) deadweight losses $D_u(t) = Y_0(0) - Y_0(t)$ and $D_r(t) = \Pi_0(0) - \Pi_0(t_r)$. If a distortion free transfer occurs, such as an inescapable Pigouvian poll tax or an unexpected lump sum transfer, then $D_u = D_r = 0$.

Let S^u and S^r represent the intensity of political support of urban and rural members, respectively. Two factors are assumed to determine a member's intensity of political support. These are *relative income* between groups and *redistributed income* within groups. Relative income is the level of per capita income relative to that of a member in the other group. Redistributed income is the difference between the within group per capita incomes before and after government intervention.

Thus members observe their neighbor's plate and compare the size of the cake they see there with what they have on their own plate. They also remember the size of the cake they had before the government redistribution. The larger their own piece of cake relative to their neighbor's piece and the larger their piece relative to the one they enjoyed prior to redistribution, the greater is their intensity to support the government.

In view of the above, the urban and rural political support functions S^u and S^r are formulated as:

$$S^u(t) = W_1 G(Y(t) - \Pi(t)) + W_2 F(Y(t) - Y(o)) \tag{3}$$

and

$$S^r(t) = W_1 G(\Pi(t) - Y(t)) + W_2 F(\Pi(t) - \Pi(o)) \qquad (4)$$

where the functions $G(\cdot)$ and $F(\cdot)$ are each non-negative, increasing and strictly concave and W_1 and W_2 are non-negative coefficients that sum to unity and are assumed to be equal for all individuals in each group.[5] The constant W_1 indicates the importance given by members to their relative incomes in supporting the government while W_2 reflects the importance of redistributed income.[6] Specifications (3) and (4) incorporate two main restrictions: (i) both S^u and S^r are additively separable in both *relative and redistributed* income; (ii) the effect of these factors is identical for all members (i.e., the same G and F functions appear in S^u and in S^r). These restrictions simplify the analysis and allow us to derive simple and illuminating results below. We leave the analysis of the general case to future research.

Let $h(t)$ indicate the *relative income* between the rural and urban sectors,

$$h(t) = \Pi(t) - Y(t) \qquad (5)$$

Thus $h(0)$ is disparity in the endowment income between the rural and urban sectors. If $h(0)$ is positive initially before redistribution, then the rural sector is better off compared to the urban sector. Therefore, $|h(0)|$ is the magnitude of the level of disparity in endowment incomes.

The model requires that no political support is forthcoming from the group that is initially at a disadvantage unless this group is made better off. This requirement is stated formally as

$$G(x) = 0 \text{ for all } x \le -|h(0)|. \qquad (6a)$$

We shall also require $G(\cdot)$ to satisfy

$$G'(x) \to \infty \text{ as } x \text{ approaches } -|h(0)| \text{ from above;} \qquad (6b)$$

this will ensure the existence of a solution to the problem below.

We note that $h(t)$ is increasing in t, i.e.,

$$h'(t) = [\Pi_0'(t_r) + 1]R - Y_0'(t) + 1 > 0 \qquad (7)$$

($\Pi_0'(t_r)$ represents $\partial \Pi_0 / \partial t_r$). Condition (7) follows from the fact that both $\Pi_0'(tR)+1$ and $1-Y_0'(t)$ are positive. Suppose $t > 0$, then both Π_0' and Y_0' are negative, but Π_0' must exceed minus unity since otherwise a rural member would become worse off by receiving a larger subsidy. Likewise when $t < 0$, both Y_0' and Π_0' are positive but Y_0' must be less than unity since otherwise the urban member would become better off by receiving a smaller (positive) transfer. We summarize these observations for future reference:

$$\Pi_0'(tR) + 1 > 0 \quad \text{and} \quad 1 - Y_0'(t) > 0. \tag{7a}$$

The government chooses t in order to maximize total political support:[7]

$$S(t) = n_u \, S^u(t) + n_r \, S^r(t). \tag{8}$$

The first order condition requires that the optimal per capita transfer t^* satisfies

$$W_1 h'(t^*)[G'(h(t^*)) - RG'(-h(t^*))] + W_2 R[F'(g_u'(t^*))g_u(t^*) \\ + F'(g_r'(t^*))g_r(t^*)] = 0 \tag{9}$$

where

$$g_u = Y(t) - Y(o) \tag{10a}$$

and

$$g_r = \Pi(t) - \Pi(o) \tag{10b}$$

are the redistributed incomes.

The transfer level t^e is of particular interest, defined from

$$h(t^e) = 0, \tag{11}$$

which yields full equity between the rural and urban sectors. From (7) t^e is unique; it is positive, zero or negative whenever $h(0)$ is negative, zero or positive. Reducing income inequality is a desirable property of any transfer policy. Transfers, however, impose an unwanted burden on the economy in the form of deadweight losses. Thus there is always the trade-off between equity and efficiency. The full equity transfer t^e therefore serves as a reference point by which different government policies are evaluated. If equity considerations dominate, then we expect $t^* = t^e$; if efficiency considerations dominate, then $t^* = 0$. As expected, and demonstrated below, the *relative income* factor introduces (via the G function) equity considerations into government transfer policies; the *redistributed income* factor incorporates (via the F function) efficiency considerations. We shall investigate each of these factors separately by considering first the extreme cases ($W_1 = 1$; $W_2 = 0$) and ($W_1 = 0$; $W_2 = 1$). The more general case in which both W_1 and W_2 are positive will then be discussed.

To a large extent, two structural parameters characterize the economy under consideration. These are R, the population ratio between the urban and rural sectors, and $h(0)$, the disparity in endowment incomes. We pay special attention to the effects of these parameters on the optimal transfer t^*.

The Pure Relative Income Effect ($W_1 = 1$; $W_2 = 0$)

In this scenario, individuals deciding on their political support are concerned only with their income relative to that of members of the other group. Condition (9) becomes:

$$G'(h(t^*)) = RG'(-h(t^*)) \qquad (12)$$

The relation between t^* and t^e is summarized in the following:

Proposition 1: $t^* =$ (resp. >, <) t^e whenever R = (resp. <, >) 1.

The proof follows directly from (12), the strict concavity of G and (7); it is demonstrated graphically in Figures 8-1 and 8-2.

Figure 8-1 considers the case $h(0) < 0$. The curve labeled $h_1(t)$ corresponds to the case R = 1; the $h_2(t)$ and $h_3(t)$ curves correspond to R > 1 and R < 1, respectively.[8] The corresponding equity transfers are t_1^e, t_2^e and t_3^e. Note that the equity transfer decreases with R($t_2^e < t_1^e$) because an increase in R implies that there are more members in the urban sector relative to the rural sector. Thus a given per capita urban tax generates a larger per capita rural subsidy.

Suppose R = 1 and assume that a transfer level $t' < t_1^e$ is chosen. This results in an income disparity level of $h(t')$. Since $G(\cdot)$ is strictly concave, $G'(\cdot)$ decreases in h and hence, as Figure 8-1 shows, $G'(h(t')) > G'(-h(t'))$, violating condition (12). Likewise, $G'(h(t)) < G'(-h(t))$ for all $t > t_1^e$. It follows therefore that condition (12) is satisfied only at the equity transfer t_1^e. In a similar manner, one can show using Figure 8-1, that R > 1 entails an optimal transfer t^* which is smaller than the equity transfer t_2^e (but positive), and when R < 1, the optimal transfer exceeds the equity level t_3^e.

When R = 1, the two sectors are equal in size; there is no reason to favor one group over the other (each vote counts the same). Hence, full equity is attained. When R > 1 so that there are more urban than rural members, the tax policy is biased toward the larger group and the transfer level is smaller than the one needed to achieve equity. When R < 1, the transfer policy favors the rural sector; the optimal transfer exceeds the equity transfer t_3^e. In this case the government "overshoots" in redistributing income such that the income of rural members after redistribution exceeds that of urban members, even though initially the urban sector is better off ($h(0) < 0$).

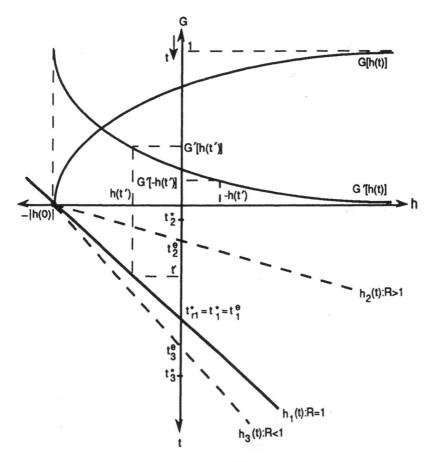

Figure 8-1. The Determination of Optimal Transfers with Endowment Incomes Higher in the Urban Sector.

The direct effect of R is summarized in:

<u>Proposition 2</u>:   (a) $dt^*/dR < 0$ provided $h(0) < 0$; (b) $dt_r^*/dR < 0$
provided $h(0) > 0$ (where $t_r^* = Rt^*$).

<u>Proof</u>:   (a) From (5), $\partial h(t^*)/\partial R = (\Pi_0' + 1)t^*$ is positive or negative
whenever t^* is positive or negative, respectively [cf. (7a)]. t^* is positive (negative) whenever $h(0)$ is negative (positive). Suppose R
changes by the increment dR. Let dh^* be the increment in $h(t^*)$ that
preserves the optimality condition (12). That is, dh^* satisfies:
$G''(\tilde{h}(t^*))dh^* = -RG''(-\tilde{h}(t^*))dh^* + dR \cdot G'(h(t^*) + dh^*)$, where $\tilde{h}(t^*)$ is
a value between $h(t^*)$ and $h(t^*) + dh^*$. Divide both sides by dR to get
$(dh^*/dR)[G''(\tilde{h}(t^*)) + RG''(-\tilde{h}(t^*))] = G'(h(t^*) + dh^*)$ which implies by
virtue of the monotonicity and strict concavity of $G(\cdot)$ that dh^*/dR is
negative. But

$$dh^*/dR = \partial h^*/\partial R + (\partial h^*/\partial t^*)(\partial t^*/\partial R),$$

where $\partial h^*/\partial R > 0$ (as discussed above) and $\partial h^*/\partial t^* > 0$ (cf. conditions
(7)). Thus it must be that $\partial t^*/\partial R$ is negative. To prove (b) note that
$\partial t_r^*/\partial R = \partial(Rt^*)/\partial R = t^* + R\partial t^*/\partial R$. By using the explicit expressions of
$\partial h^*/\partial R$ and $\partial h^*/\partial t^*$ in the equation above, we obtain

$$\partial t^*/\partial R = [(\Pi_0' + 1)R + 1 - Y_0']^{-1}(-(\Pi_0' + 1)t^* + dh^*/dR).$$

Thus,

$$\partial t_r^*/\partial R = t^*\left\{1 - [(\Pi_0'+1)R+1-Y_0']^{-1}(\Pi_0'+1)R\right\}$$
$$+ [(\Pi_0'+1)R+1-Y_0']^{-1}dh^*/dR$$

The second term is negative, since dh^*/dR is negative (see above) and
the bracketed term is positive (follows from (7a)). The term inside the
curly bracket is positive since $1-Y_0' > 0$ ((7a) again). Noting that
$h(0) > 0$ implies $t^* < 0$ completes the proof.
    Thus, when $h(0) < 0$, the optimal urban tax t^* decreases as a result of
an increase in the number of urban members relative to the rural sector.
As R increases, (1) there are more members to tax (or relatively fewer
members to subsidize); and (2) the urban group becomes relatively more

important to the government in terms of political support. These two effects reinforce each other such that t^* declines with R.

On the other hand, the per capita transfer to the rural sector $t^*_r = t^*R$ can increase or decrease because $\partial t^*_r / \partial R = t^*(\xi_{R+1})$ and $\xi_R \equiv (\partial t^*/\partial R)(R/t^*)$ may be smaller than minus unity. The impact of a change in R on t^*_r is indeterminate because there are two offsetting effects. First, there are relatively fewer members in the rural sector to be subsidized as R increases. This income effect results in the per capita subsidy t^*_r to increase. Second, a smaller rural population becomes relatively less important to the government in terms of political support and so transfers to the rural sector would tend to decrease. This latter substitution effect is captured by ξ_R which decreases in absolute value as R increases. Hence, it is more likely that dt^*_r/dR is positive with larger values of R.

In industrial countries, where the number of farmers relative to the urban population is declining, it is possible that per capita transfers are increasing while the per capita tax on the urban sector continuously declines. This occurs independent of the ability of an interest group to organize and pressure politicians as argued by Olson (1985, 1986). Hence, it is possible in this model to observe countries with relatively fewer farmers having higher per capita transfers. It is in the interest of support maximizing politicians to do so even though there are relatively fewer farmers to obtain votes from.

Figure 8-2 depicts the case $h(0) > 0$. The value of t^* is negative, i.e., the government transfers income from the rural to the urban sector (as is widely observed in developing countries.) As in the previous case, the optimal transfer coincides with the full equity transfer when R equals unity. The optimal transfer is less than the full equity transfer when $R > 1$ and is greater than the full equity transfer when $R < 1$.

Variations in endowment incomes due to changes in technology or input costs (in either the agricultural or industrial sector) shift the h(t) function in Figure 8-1. For example, if the productivity in the industrial (urban) sector improves relative to the agricultural (rural) sector, then Y_0 increases relative to Π_0. This results in a leftward shift of the h(t) function and an increase in the optimal urban tax t^* or rural subsidy t^*_r. Conversely, technological change in the agricultural sector that raises the relative productivity and endowment incomes of the rural sector, ceteris paribus, will result in the function h(t) to shift right in Figure 8-1 so that t^* decreases. This result has been prevalent in North America and Australasia where government supported research and extension has resulted in large technological advances.

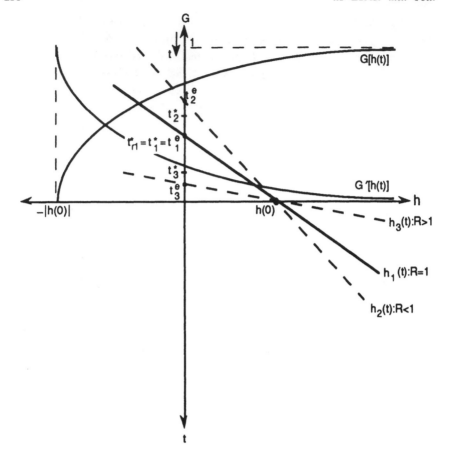

Figure 8-2. The Determination of Optimal Transfers with Endowment
Incomes Higher in the Rural Sector.

This may partially explain Anderson and Tyers observation that protection to farmers is relatively lower in these countries.

The insistence by West Germany (FRG) for higher cereal support
prices while the United Kingdom (UK) is most strongly opposed can
also be partially explained by the relative difference in endowment
incomes between the rural and urban sectors. The FRG has many more
cereal farms than the UK and are smaller in size and inefficient. From
a strict interest group point of view, one would expect the UK to favor
higher cereal prices. However, the increases in productivity in the industrial sector in FRG have exceeded that of the UK in past decades
while the reverse is true in agriculture. Hence, the relative propensity
of political preferences in price policy in Europe can be explained in

part by the divergence in relative endowment incomes between agriculture and manufacturing across countries. A similar argument could be made to explain the price policy differences between Canada and Argentina. As both countries are highly urbanized societies with similar agro-climatic conditions, Argentina taxes farmers while the reverse is the case in Canada. We summarize this critical impact of endowment incomes in the following proposition, which can be verified using Figures 8-1 and 8-2.

Proposition 3: The government's optimal choice of t^* decreases with $h(0)$.

While $h(0)$ is the intercept of the $h(t)$ function, the location and shape of the $h(t)$ depends also on its slope. This slope depends on R and on the marginal deadweight losses Π_0' and Y_0'. The effect of R was studied above. We concentrate now on the latter.

Suppose the government changes the instruments it uses to tax and subsidize such that the urban sector is unaffected but the distortionary effects on the rural sector are more severe such that $|\Pi_0'|$ increases for all $t \neq 0$. This has the effect of reducing rural income for a given level of transfer and causes $h(t)$ to turn clockwise (counter clockwise) about the point $h(0)$ when t is positive (negative). It can be determined from Figure 8-1 that, when $h(0)$ is negative, the equity transfer t^e will increase and so will the optimal transfer t^*. When $h(0)$ is positive, t^e is negative and will increase (become less negative) as a result of a counter clockwise rotation of $h(t)$ about $h(0)$. It can be shown that t^* increases as well. An (ceteris paribus) increase in $|Y_0'|$ for all $t \neq 0$, will have an opposite effect.

These results have important implications for observed patterns of government intervention in agricultural markets. Commodity sectors with more elastic supply functions tend to make the producers (the rural sector) more vulnerable to distortionary forces. Hence, one would expect higher producer subsidies (or lower producer taxes) in such cases. On the other hand, inelastic demand decreases deadweight loss and tends to decrease $|Y_0'|$ as well. Thus, one would expect higher transfers in those sectors. USDA calculations indicate that inelastic demand commodities like fluid milk and wheat have higher "producer subsidy equivalents" world-wide than more elastic commodities like meat and feed grains. Furthermore, the efficiency of transfers are higher for importers than for exporters. It is generally recognized that importers protect agriculture more than exporters in industrial countries because the latter's terms of trade decline (improve) with the subsidization (taxation) of the rural sector (and vice-versa for importers). This may partially explain Anderson and Tyers' observation that North America and Australasia subsidize agriculture less because they are predominantly exporters and hence limit subsidies. On the other hand, the dairy and sugar sectors in the United States as net importers have higher rates of protection than export sectors (USDA).

The Pure Redistributed Incomes Effect ($W_1 = 0$; $W_2 = 1$)

In this scenario, individuals are only concerned with their income with government intervention relative to that with no government intervention, or endowment income. The necessary condition (11) becomes

$$F'(g_r(t^*))(\Pi_0'(t_r^*) + 1) = F'(g_u(t^*))(1 - Y_0'(t^*)), \qquad (13)$$

from which the following result follows.

Proposition 4: If a member's political support depends only on the change in income due to redistribution, then the optimal transfer is zero.

Proof: From $\Pi_0'(0) = Y_0'(0) = g_r(0) = g_u(0) = 0$, it follows that $t^* = 0$ satisfies (13). Furthermore, $t^* = 0$ is the unique solution to (13), for suppose that t^* increases from zero. Then the left hand side of (13) decreases because (i) g_r increases (becomes positive) which, by virtue of the strict concavity of F, causes $F'(g_r)$ to decrease, and (ii) Π_0' becomes negative so that $\Pi_0'+1$ decreases. Likewise, both g_u and Y_0' become negative, which causes the right hand side of (13) to increase. Thus $t^* > 0$ cannot be a solution. In a similar manner, a negative transfer is not possible, leaving $t^* = 0$ as the unique solution.

The implication of Proposition (4) is that governments will not redistribute income when maximizing political support if redistributed income is the only factor affecting individuals, behavior in supporting the government. This result holds regardless of the relative group size, the extent of income inequality between the groups and of the marginal deadweight loss of redistribution.

The Mixed Case ($W_1 > 0$; $W_2 > 0$)

If both relative and redistributed incomes affect political support, then the tendency for governments to reduce the disparity in income distribution is partially mitigated by the effect of redistributed income on the level of political support. To show this, consider $R = 1$ in condition

(11). If $t^* = t^e$, then the first term on the left-hand side of (9) is zero but the second term does not vanish. Hence, full equality with $R = 1$ cannot be optimal. The term associated with W_2 will push the optimal transfer toward zero.

Hence, economic efficiency considerations dampen the government's propensity to redistribute income away from relatively advantaged groups. The extent to which this occurs depends critically on the relative value of the weights W_1 and W_2.

Some More Evidence

The above results suggest that a country that subsidizes farmers is likely to have lower per capita endowment incomes in the rural sector than in the urban sector (and vice-versa for countries that tax agriculture). This is consistent with the observations by Anderson and Hayami (1986) that countries who subsidize farmers have high productivity rates and wages in the manufacturing sector (where one can hypothesize that rural endowment incomes are relatively lower) while the opposite is the case in developing countries or in times before industrialization as in the case of Japan and Europe when agriculture was taxed. Hence, one should not be surprised that wheat prices are the highest in Saudi Arabia (Byerlee and Sain 1986) and the lowest in Ethiopia (Kerr 1985).

Furthermore, Honma and Hayami (1986) find a statistical correlation between agricultural protection and its comparative advantage vis-a-vis the manufacturing sector. Hence, agriculture may have a comparative advantage in many developing countries so that endowment incomes would be higher and hence the rural sector is taxed. In addition, export crop sectors in developing countries have a comparative advantage over import-competing food crop sectors such that the latter are taxed less (Kerr; Krueger, Schiff and Valdes 1988). This occurs even though farmers in the export sector in developing countries are typically fewer and more able to organize as a pressure group.

The observation by Anderson and Tyers (1988) that Australasia and North America, although highly industrialized economies, tend to protect farmers less may in part be explained by the fact that relative endowment incomes in agriculture may not be so low as in other countries because of agro-climatic and the advancements in technology due to publicly funded research.

A fascinating feature of European agricultural politics has been the increasing polarization between the UK and the FRG on matters relating to support prices, particularly in regard to cereal prices. The UK has argued for lower price supports while the FRG has been the most vocal for the opposite. This occurs even though the total costs of the cereals regime is higher for the FRG because (a) FRG's taxpayer contributions are higher because the value-added tax is based on the level of GNP and the UK is one of the lower income members and (b) FRG's consumer costs are higher because they are a net importer of cereals (unlike the UK) resulting in invisible transfers resulting from intra-EC trade at supported prices while contributing more to importer levy income resulting from off-shore trade. Furthermore, the benefits of the cereal regime is greater for the UK because of larger production such that the level and per capita farm benefits are higher in the UK. Hence, it appears puzzling why the UK is so adamant for lower prices and why the FRG argues for the opposite. This irony is further complicated by the fact that in the UK, farms are larger and far fewer in number, leading the Olson-Becker interest group models to predict that the price preferences of the UK and the FRG would be reversed.

Olson (1985, 1986) argues that larger and fewer firms promote the political power of an industry by reducing costs of organizing, preventing free-riding and mitigating opposition. Olson (1985) and Gardner also argue that a higher variability in farm size and a lower geographic dispersion of farms would lead to more lobbying and higher subsidies. Yet the farms in the FRG are far more uniform in size and are more geographically dispersed, evidence that is contrary to the revealed political preferences for prices by the UK and the FRG and hence contradicts the predictions of interest-group models.

The model developed in this paper emphasizes the importance of the relative rural-urban endowment income differential in order to explain UK and FRG relative price preferences. The UK has few, large and cost-efficient farms while in the FRG farms are many, small and high cost. On the other hand, the industrial sector in the FRG is very rich while the UK's manufacturing sector has lagged far behind in terms of productivity and wages in the post-war era. Hence, the pre-policy income gap between the rural and urban sectors are very high in the FRG while the opposite is the case in the UK. We argue that this pre-policy disparity in relative incomes is a fundamental force in the current and historical political economy of agricultural policy not only in Europe but also world-wide.

Conclusion

This paper develops a theoretical model to explain the observed patterns of government intervention in agriculture. A model is developed to explain Anderson and Tyer's observation that society has an income elastic demand for assisting farmers. Politicians maximize political support from members of the urban and rural sectors. These support functions are assumed to depend on two factors: relative income between groups and redistributed income within groups. Politicians exploit the differences in the marginal utility of income resulting from differing endowments such that political support is maximized with redistribution. Government policies in agriculture are viewed as an outcome of political support maximization in dealing with societal income distribution.

The results show that per capita transfers depend critically on relative endowment (pre-policy) incomes and numbers in each group. Conditions are derived to determine when farmers are taxed or subsidized, and when over-shooting occurs, i.e., when transfers result in a switch in the relative incomes between groups. Because political support is also a function of the change in income within groups due to redistribution, the tendency for governments to reduce the disparity in relative incomes is partially mitigated. It is also determined that the marginal deadweight loss of redistribution affects the equilibrium transfer.

A generalization of the model developed in this paper could proceed along several lines of research. The weights W_1 and W_2 may be endogenized by specifying them to vary over time or across commodity sectors and countries. In addition, the weights may vary between the urban and rural sectors. Our model also assumes that the transformation of economic welfare into political support, i.e., the functions $G(\cdot)$ and $F(\cdot)$, is the same over time and between sectors, groups and countries. However, Downs emphasizes the importance of lower information and uncertainty costs for these groups whose welfare is affected more by the transfer policy. The implication for the model developed in this paper is that $G(\cdot)$ and $F(\cdot)$ should be specified to be different between groups, policies, sectors and countries.

A further potential modification to the model is the specification of an explicit preference function for politicians (governments). This would allow for the effect of socio-political characteristics of politicians such as ideology and for differing political regimes (e.g. democratic vs. authoritarian, etc.). Another critical aspect is the effect of the political system of representation in transforming the economic basis of the policy costs/benefits into political costs/benefits. For example, representation based on geography can play a special role and so equation (8) could be modified to reflect the effect of "distributive politics" (Weingast, Shepsle and Johnsen 1981). A parliamentary system in the UK compared to proportional representation in the FRG may partially explain the relative success of farmers in the FRG.

This paper focuses on income transfer policies in agriculture but ignores the importance of "public good" type policies that, although increase social welfare, may have important implications on the distribution of income between groups and hence on the outcome and the choice of income transfer policy (de Gorter 1983 and de Gorter and Zilberman 1990). This paper also ignores the issue of policy instrument choice, a discrete choice that is interrelated and simultaneously determined by the level of income transfers (Rausser and de Gorter 1989). Finally, the model in this paper can be extended to include the effects of lobbying and pressure group activities on the equilibrium transfer. The Olson-Becker interest group models adequately determine the factors affecting the ability of groups to organize and pressure politicians. The political support functions would have to be augmented to include the effects of these pressure activities. Some evidence in agriculture on how pressure group activities affect politician behavior in setting price supports is given in de Gorter and Rausser (1989).

Notes

1. A plethora of studies give evidence, a few of which include Schultz, Peterson, Bale and Lutz, Binswanger and Scandizzo, World Bank, OECD, USDA, Kerr, Anderson and Hayami, and Krueger, Schiff and Valdes.

2. Exploitation of agriculture in developing countries to finance industrialization has been a justification for economic development. See, for example, Johnston and Mellor.

3. See, for example, de Janvry.

4. For applications to agriculture, see Olson (1985, 1986), Becker, Gardner, Bullock, Bates and Rogerson, and Balisacan and Roumasset.

5. Although assumed exogenous in our analysis, these weights, W_1, W_2, may in fact vary across socio-economic characteristics for individuals within groups or between groups. Indeed, Downs emphasized the latter in showing that uncertainty and differential information affected voter preferences and government policies. Voters are well informed on issues affecting them as income earners, giving rise to inequality in political influence in favor of producers. The latter have lower information and uncertainty costs because benefits are concentrated and hence producers are more sensitive in their political support.

6. The W_1 weights may be a function of endowment incomes themselves. For example, the poorer one is, the more likely they are concerned about their own well-being relative to no government intervention. Hence, the value of the weight W_2 in equations (5) and (6) could be higher in poor countries relative to rich countries. However, we assume in this paper that the weights are fixed and independent of endowment incomes.

7. Equation (8) assumes politicians maximize the sum of individual support functions. However political support is not simply additive because in reality political institutions transform the economic basis of policy costs/benefits into political costs/benefits. An example is distributive politics or disproportionate representation in the United States whereby each state is represented by two senators. For example, the wheat sector is likely to receive higher subsidies because states like Kansas have few consumers and other major producing sectors for which wheat farmers would have to compete with in having their senators support outcomes in their favor. Political districting and cost accounting as developed in Weingast, Shepsle and Johnsen could be applied in an appropriate manner to equation (8). The "political influence" of voters not only varies across sectors within a country but also across countries where differing political institutions could result in different policy outcomes.

8. The $h(t)$ functions are shown to be linear which is the case only when the marginal deadweight loss of the transfer is zero. It is not possible to determine the exact shape of $h(t)$ a priori except under strict conditions.

References

Anderson, K. and Y. Hayami eds. 1986. *The Political Economy of Agricultural Protection: East Asia in International Perspective*. Sydney: Allen and Unwin.

Anderson, K. and R. Tyers. 1988. Agricultural Protection Growth in Advanced and Newly Industrialized Countries. *Agriculture and Governments in an Interdependent World*. A. Maunder ed. Aldershot: Gower, forthcoming.

Bale, M. D. and E. Lutz. 1981. Price Distortions in Agriculture and Their Effects: An International Comparison. *American Journal of Agricultural Economics*. 63:8-22.

Balisacan, A. and J. Roumasset. 1987. Public Choice of Economic Policy: The Growth of Agricultural Protection. *Weltwirtschaftliches Archiv*. 123:232-47.

Bates, R. and W.P. Rogerson. 1980. Agriculture in Development: A Coalitional Analysis. *Public Choice*. 35:513-27.

Becker, G. 1983. A Theory of Competition Among Pressure Groups for Political Influence. *Quarterly Journal of Economics*. 98:371-400.

Binswanger, H. and P. Scandizzo. 1983. *Patterns of Agricultural Protection*. Report ARU 15. Washington, DC: World Bank.

Breton, A. 1974. *The Economic Theory of Representative Government*. Chicago: Aldine Publishing Co.

Bullock, D.S. 1988. *The Volatility of Farm Program Transfers: A Political Pressure Group Approach*. Unpublished paper, Department of Economics, University of Chicago.

Byerlee, D. and G. Sain. 1986. Food Pricing Policy in Developing Countries: Bias Against Agriculture or for Urban Consumers? *American Journal of Agricultural Economics*. 68:961-69.

de Gorter, H. 1983. *Agricultural Policies: A Study in Political Economy*. Ph.D. Thesis, Department of Agricultural and Resource Economics, University of California, Berkeley.

de Gorter, H. and G.C. Rausser. 1989. *Endogenizing U.S. Milk Price Supports*. Giannini Foundation Working Paper #504. Department of Agricultural and Resource Economics, University of California, Berkeley.

de Gorter, H. and D. Zilberman. 1990. On the Political Economy of Public Good Inputs in Agriculture. *American Journal of Agricultural Economics*. 72:131-37.

de Janvry, A. 1983. Why Do Governments Do What They Do? The Case of Food Price Policy. *The Role of Markets in the World Food Economy*. D. G. Johnson and E. Schuh, eds. Boulder, Colorado: Westview Press.

Downs, A. 1957. *An Economic Theory of Democracy*. New York: Harper and Row.

Frey, B.S. and L.J. Lau. 1968. Towards a Mathematical Model of Government Behavior. *Zeitschrift fur Nationalokonomie*. 28:355-80.

Gardner, B.L. 1987. Causes of U.S. Farm Commodity Programs. *Journal of Political Economy*. 95:290-310.

Hayami, Y. 1988. *Japanese Agriculture Under Siege, The Political Economy of Agricultural Policies*. London: Macmillan.

Honma, M. and Y. Hayami. 1986. Structure of Agricultural Protection in Industrial Countries. *Journal of International Economics*. 20:115-129.

Johnston, B.F. and J.W. Mellor. 1961. Agriculture in Economic Development. *American Economic Review*. Vol. LI: 566-93.

Kerr, T.C. 1985. *Trends in Agricultural Price Protection: 1967-83*. Background Working Paper for the 1986 World Bank Development Report.

Krueger, A., M. Schiff and A. Valdes. 1988. Agricultural Incentives in Developing Countries: Measuring the Effect of Sectoral and Economywide Policies. *World Bank Economic Review*. 2:255-72.

OECD. 1987. *National Policies and Agricultural Trade.* Paris: Organisation for Economic Cooperation and Development, May.

Olson, M. 1985. Space, Organization and Agriculture. *American Journal of Agricultural Economics.* 67:928-37.

————. 1986. The Exploitation and Subsidization of Agriculture in the Developed and Developing Countries. *Agriculture in a Turbulent World Economy.* A. Maunder and U. Renborg eds., pp. 49-59 . Aldershot: Gower Publishing Group.

Peltzman, S. 1976. Toward a More General Theory of Regulation. *Journal of Law and Economics.* 19:211-40.

Peterson, W.L. 1979. International Farm Prices and the Social Cost of Cheap Food Policies. *American Journal of Agricultural Economics.* 61:12-21.

Rausser, G.C. and H. de Gorter. 1990. Endogenizing Policy in Models of Agricultural Markets. in *Agriculture and Governments in an Interdependent World.* A. Maunder, Ed. Aldershot: Gower Publishing Group.

Schuh, G.E. 1978. Approaches to Basic Needs and Equity that Distort Incentives in Agriculture. *Distortions of Agricultural Incentives.* T.W. Schultz, ed. Bloomington: Indiana University Press.

Schultz, T.W. ed. 1978. *Distortions of Agricultural Incentives.* Bloomington: Indiana University Press.

USDA. 1988. *Estimates of Producer and Consumer Subsidy Equivalents.* ERS, Agriculture and Trade Analysis Division.

Weingast, B.R., K.A. Shepsle and C. Johnsen. 1981. The Political Economy of Benefits and Costs: A Neoclassical Approach to Distributive Politics. *Journal of Political Economy.* 89:642-65.

World Bank. 1986. *World Development Report.* Oxford University Press, Chapters 4-6.

Discussion

Masayoshi Honma

Dr. de Gorter and Dr. Tsur developed a formal model trying to explain the patterns of agricultural policies. I found their paper interesting and enjoyed reading it. The model they developed is relatively simple in structure and it makes the points to be discussed clear.

They assume two groups in an economy: urban and rural. The government in their model seeks an income transfer so that the government maximizes the total political support from the two groups. The people or voters are supposed to change their political support to the government as the changes in two factors occur; one factor is relative income between the groups and the other is redistributed income within a group. Dr. de Gorter and Dr. Tsur examine effects of relative income and redistributed income separately. We may call the first case a "relative income hypotheses" and the second "a redistributed income hypothesis." I will discuss each hypothesis separately and finally give my comment on the applicability of their model to the real world.

Let me start to discuss the relative income hypothesis first. This is the main part of their paper and most of the results and implications draw on this hypothesis. It is assumed that urban and rural people share the same political support function, which is a function of income disparity, but their positions on the political support curve are different unless their incomes are the same. Namely, the marginal intention to support the government is different by group when there is an income gap initially. Therefore, the government can increase the political support by exploiting the differences in marginal utility of income. The income transfer thus occurs from the rich to the poor. The level of transfer is determined so as to equalize the weighted marginal utilities of income of both groups; the weight is the number of people or voters in each group.

As long as the marginal utility of income is a monotonically decreasing function of income, transfer is always from the rich to the poor. This is a crucial point of their model. It is also concluded that the larger the poor is in population, the larger the transfer is. Furthermore, if the poor people are dominant in population, the transfer will be "over-shot" and the income of initially poor people will exceed that of initially rich people. These conclusions depend on the assumption that all the people are politically homogeneous in response to income changes sharing the same political support function. This view of the political market for agricultural policy seems to be too naive to explain what is happening in the real world.

Let's take a look at developing countries, for example. In most of developing countries, governments adopt policies taxing agriculture and rural people who are poor and dominant in population. If de Gorter and Tsur's model is applied to them, the rural people must be subsidized rather than taxed, and even become richer than the urban people after income transfer. The reality is opposite to what their model suggests. Dr. de Gorter and Dr. Tsur tried to explain the taxation policy on agriculture in developing countries by assuming that the initial income, or endowment income in their term, of rural people is higher than that of urban people in those countries. But it is hard to accept such assumption. If rural people are better off initially, there is no reason for them to migrate to urban area. Rural-to-urban migration, however, is commonly observed in most developing countries regardless of level of taxation. Rural people in developing countries are taxed not because they are rich but because they are politically also poor. Taxation on agriculture in developing countries should be explained in different manner.

By the way, Dr. de Gorter and Dr. Tsur referred to comparative advantage for the initial income. But the two concepts are different. A country may have comparative advantage in agriculture even if agricultural income is initially behind industrial income because comparative advantage is determined in a comparison of economic conditions with other countries.

In the second hypothesis on voters' behavior, voters are assumed to respond only to redistributed income within a group. Namely, the intensity of political support in this redistributed income hypothesis is specified as a function of income differences within each group between the initial income and the income after transfer. This specification seems to me somewhat *ad hoc* because the zero solution of optimal transfer is obvious under the monotonically decreasing marginal utility of income. In this specification, before transfer everybody shows the same intensity of political support because redistributed income is zero for all and the political intensity function is common to all. As far as the political intensity function is strictly concave, the resistance against income transfer by those who are taxed is always stronger than the support to the transfer by those who are subsidized. This means that any income transfer results in a net loss of total political support to the government. Therefore, the government has no incentive to introduce income transfer under this hypothesis.

This happens because the model assumes that one dollar loss causes the same pain to any loser regardless of his income level. But we usually consider that rich people have less pain than poor people by one dollar loss as far as marginal utility of income is a decreasing function of income. Therefore, the variable in the intensity function under the redistributed income hypothesis should be such a ratio of income transfer to the initial income as reflects the differences in importance of transfer by income level.

Additionally, Dr. de Gorter and Dr. Tsur discussed the reluctancy of the government to transfer as a result of economic efficiency considerations. But the redistributed income hypothesis is not relevant to discussion of efficiency because the zero solution of optimal transfer does not depend on the deadweight loss by transfer at all. The government is reluctant to income transfer not because of the deadweight loss but because of the strong resistance by loser.

Finally, my comment turns to the applications of their model to the real world. I already discussed the difficulty to apply their model to developing countries because the model can not explain well the taxation policy on agriculture. However, the model was found useful to explain the experiences of the United Kingdom (U.K.) and West Germany (W.G.). The difference in attitude of agricultural policy between the two countries is attributed to the difference in the initial income gap between them. W.G. has larger income gap which results in higher level of agricultural protection than the U.K. Also, the relative population contributes to explaining the difference between the two because the U.K. has larger urban population ratio, which leads to lower level of agricultural protection in the model, through this point is not mentioned in their paper.

The model developed in the paper seems to fit some cases in developed countries, especially in those who are in the phase of economic development from a rapidly industrializing economy to a more mature one. The main source of agricultural protection is the income discrepancy against agriculture combined with the difficulty of intersectoral resource reallocation in the process of rapid economic development. However, if a country is economically matured and the industrial adjustment is once completed there is no need to protect agriculture for reasons of income gap because the income gap is disappearing. Thus, the model in the paper is consistent with agricultural policies in this late stage of economic development but not relevant to those in the earlier stage.

One of the possible revisions of their model to explain the taxation on agriculture in the early stage of development is incorporating with the so-called interest group approach. Dr. de Gorter and Dr. Tsur insisted a contrast of their model to that of M. Olson, one of the representatives of the interest group approach. But the two models are not inconsistent and can be rather incorporated. Olson's idea of the "collective action" helps to understand why rural people in developing countries are taxed despite their majority in population. When a group is large in number, the free rider problem in collective action prevents them from acting efficiently in political lobbying. Such free rider problem diminishes as the group becomes small and the efficiency of political action increases. But it does not monotonically increase and there must be a threshold point in size beyond which the political power of the group becomes declining. This may apply to the U.K. whose agriculture is considered having passed the threshold point in size already.

The model developed by Dr. de Gorter and Tsur presents a basic idea of political inference which can be extended to various directions in agricultural policy analysis. Agricultural policy research is still at an infant stage of development, but vigorous efforts are being made in this field. Their model is considered one of them.

9

Summary

Stephen L. Haley and Jerry Sharples

Trade economists are concerned with two fundamental issues: Why does trade between nations take place, and when it does, what determines the commodity composition of that trade? As the chapters in this book attest, resource-based trade theories (like Heckscher-Ohlin) are no longer the only relevant theories for explaining the direction and magnitude of existing trade flows. Krugman (1981) argues that countries with similar factor endowments trade because of scale economies, and most of that is imperfectly competitive intra-industry trade. Firms with market power play strategic games. In Part One of this book these concepts are advanced and applied to agricultural trade.

Trade economists also are interested in what determines trade policy. The new theory of political economy starts from the proposition that the politics of policy formation are dominated by distribution rather than efficiency arguments (Krugman, 1988). Policy is the end result of the wielding of political power by interest groups. The economics of politics explains trade policy. These concepts and extensions to agricultural trade, are the focus of Part Two.

A distinguished group of economists contributed to this book. Each of them has contributed directly to the new developments in trade and political economy theory. The remainder of this summary reviews their main points and draws implications for future agricultural trade research.

New Trade Theory

New Trade Theory and Applied Studies

Richardson accomplishes two important objectives in his chapter. First, he skillfully presents the analytical framework in which gains from reduced protectionism and greater economic integration are evaluated in the imperfectly competitive setting of large and interdependent firms. Second, he surveys empirical research from calibration and econometric methodologies in measuring the gains from trade liberalization.

For a country which is a net importer of a product produced domestically in a setting of imperfect competition, trade liberalization can improve welfare in three ways. First, more open trade can reduce distortionary pricing above marginal costs. Trade increases the number of firms from which to purchase similar products. Increased competition limits the ability of any one firm to price above marginal cost. Second, increased trade can help eliminate the wasteful duplication of facilities and/or firms whose high fixed costs drive up consumer prices. Trade rationalizes a domestic industry: it forces the exit of excessive firms that drive up average costs. Third, more open trade reduces exploitative income transfers to foreign firms.

If the country is a net exporter of a product, then there exists the possibility of welfare loss in the trade setting. For instance, mark-up pricing of a domestic firm in a foreign market may be capturing benefits like those associated with an optimal tariff. If trade were liberalized, then the gain would dissipate. It may also be the case that two dominant producers in the export market can deter entry by a foreign competitor. In a liberalized setting, the entry of a foreign competitor could reduce the firms' profits, and thereby reduce national welfare. Although increased trade should increase product variety available to consumers and downstream firms, it is theoretically possible that there will be reduced varieties of domestic products caused by the exit of domestic firms.

It is therefore an empirical matter whether liberalized trade increases national welfare. Richardson reviews a number of empirical studies which evaluate the gains to trade in the setting of imperfect competition. Overall, he concludes that trade liberalization has strong positive effects on welfare. The gains are anywhere from two to three times the gains calculated under the setting of perfect competition. Trade enhances the "workability" of competition. The key assumption behind these results is that of free entry and exit of firms within an industry. Trade liberalization leads to the rationalization of industrial structure and heightened market competitiveness. Increased import competition disciplines the power of domestic firms to price above marginal cost. The competition leads to more efficient scale economies and productivity enhancement. Increased product variety benefits not only the final use consumer, but also the user of intermediate products. Conclusions regarding short-term trade-induced adjustment costs under imperfect competition are ambiguous, however. Results depend on the methodology used. Calibration methods show moderate to heavy adjustment costs, while regression methods show insignificant relationships between trade penetration and entry, exit, and concentration.

Strategic Trade Policy

Krishna and Thursby provide a selected review of the imperfect competition literature. Their selection criteria are based, appropriately enough, on what they feel is particularly relevant for agricultural trade and policy analysis.

Imperfectly competitive markets are characterized by "large" firms, whose actions affect markets and the actions of other agents. In agriculture, large grain trading firms dominate U.S. agricultural exports. In other countries, marketing boards and state trading agencies play dominant roles in both marketing and trade. These market relationships are bound tightly to the existence of increasing returns to scale in the transportation and processing of agricultural commodities, if not to the actual production of the primary product.

Krishna and Thursby present analyses of strategic policy choices, assuming particular types of policies and market structures. Policies operate either directly through quantitative restrictions, or indirectly through taxes and subsidies. The treatment of market structure involves the simple case of the monopolist, and works upward through the increasing complexity of oligopolistic relationships.

There seems to be very few results which generalize well. Analyzing how an equilibrium is altered by government policies is dependent on the model chosen for the analysis. Crucial issues include: timing (for instance, who moves first?); the strategic variable used by the firm (quantity or price?); characteristics of market structure (monopoly, duopoly, or oligopoly?; is there freedom of entry and exit?); and the set of instruments available to the government. One result that tends to be robust to a variety of specifications of market structure is that of the export restraint (ER). The ER need not be set at restrictive levels in order to have significant effects. Nonbinding ER's raise domestic prices and encourage collusive behavior of foreign and domestic firms.

Krishna and Thursby conclude that all sorts of government policy options can be appropriate depending on the relevant model. They note that the results of calibration models are intimately influenced by model specification. This caution is particularly relevant to the method of conjectural variation. Although conjectural variation parameters can measure the degree of competition or collusion in market conduct, their usefulness in policy analysis is questionable. Any inferences involving timing, which are crucial to policy analysis, are unclear. More generally, because calibration methods can only incorporate hypotheses regarding market conduct and cannot directly test those hypotheses, there is much room for guidance in model specification from econometric work.

Although commenting on Richardson's chapter, Paarlberg introduces considerable doubt on the ability to extend empirical research along the lines suggested by Krishna and Thursby. He bases his comments on the potential for analyzing the behavior of international grain trading firms. He echoes the caution that imperfect competition models are very sensitive to a number of factors. In his example, these factors include the choice of a payoff function, the reaction function of rivals, the choice of the strategic variable, the existence of capacity constraints in supplying the market, and the symmetry of trading firms. He argues that there is a premium on a detailed knowledge of the industry. However, since even the share of U.S. agricultural exports by the five largest trading firms is uncertain, the probability of obtaining proprietary data necessary to model this imperfect market is hardly promising.

Nonetheless, Thursby and Thursby present a model of U.S. and Canadian competition for the Japanese wheat market. In the model, symmetric U.S. export firms and the Canadian Wheat Board all compete against each other for Japanese market share. Thursby and Thursby calibrate the model with market data and derive conjectural variation parameters which are consistent with the data. They provide standard deviations for the parameters derived from the estimated model functions. Their results support the hypothesis of Bertrand (or price) competition among exporters in the Japanese wheat market.

The Thursby and Thursby model represents one of the few empirical attempts to apply the new imperfect competition approach to an agricultural trade issue. As with most applied work in imperfect competition, it is not immune to criticism. In her review, Veeman questions use of the data used by Thursby and Thursby, arguing that inappropriate data in the study constrains severely the confidence that can be placed in their results. Even more serious, however, is Veeman's contention that the Japanese importing agency is best modeled as a monopsonist, playing off both the U.S. firms and the Canadian Wheat Board. The political dimension is missing as well. This type of criticism underlines the need for better data (if possible), and for more econometric testing of hypothesized market relationships.

Implications for Trade Modeling

MacLaren is concerned with modeling agricultural trade of goods that are imperfect substitutes. He notes that the Armington model, although useful for modeling imperfect substitutes, has serious shortcomings. MacLaren looks to the new trade theory to provide an alternative to, or perhaps justification for the use of, the Armington specification.

The new trade theory places its emphasis on consumer preferences as an explanation for intra-industry trade patterns. Increased access to a variety of goods that would be unavailable without trade increases the utility of the representative consumer. Demand expansion resulting from increased trade allows firms to expand production out along an increasing returns to scale production schedule. Therefore, with trade, the consumer can purchase more varieties of a good at a lower cost.

Firms create variety by differentiating their products. MacLaren reviews three types of product differentiation that are discussed in recent literature. The first is horizontal differentiation, which emphasizes a consumer's desire for a variety of goods of equal quality. The second is vertical differentiation, which emphasizes quality differences. The third is technological differentiation, which emphasizes firms' preferences for intermediate goods with differing embodied technologies.

MacLaren suggests that the connection between vertical differentiation and country resource characteristics may provide the theoretical grounding of the Armington approach in agricultural trade analysis. Goods can be differentiated on the basis of contributions derived from country-specific resources. In agriculture, quality often depends on the growing environment and breed characteristics over which limited control can be exercised because of location and climate. The possible uniqueness of a country's variety may then distinguish it from that of other countries' varieties.

In her review of MacLaren's chapter, Ballenger notes that agricultural commodity analysts tend to emphasize trade rigidities rather than substitution possibilities. There are several obvious examples where preferences are important -- chicken: parts or whole; rice: japonica or indica; beef: grass-fed or grain-fed; pork: certain cuts.

Understanding these types of preferences help guide one's expectations regarding consumer reactions to product price changes. An implication is that expanding U.S. agricultural exports may involve policies that emphasize product market development over the granting of price subsidies or bonus payments.

MacLaren ends by noting the large gap between the new trade theory and empirical models. With the lack of empirical modeling, it is difficult to assess the new theories even within their presumed industrial structures, let alone agricultural trade.

Political Economy

The New Theory

Moore provides a very useful review of recent political economy models, and makes several suggestions for where he believes future work should proceed. Existing political economy models can be classified as either social concern models or self-interest models. Social concern models vary in their perspectives. On one hand, conservative status quo models emphasize the protection of existing private property rights and income distribution. Closely related adjustment assistance models deal with avoiding income losses. Opposing these models are social change models where equity concerns are of primary importance. The self-interest models emphasize either pressure group effects of lobbyists or the power of the general public ("adding machine" models).

Most recent political economy models have focused on political lobbying in the formation of trade policy. They run counter to the perspective of economists who analyze policy alternatives in terms of the theory of the second best. The new approach emphasizes the characteristics of distributional coalitions in the policy formation process, and how vote-seeking politicians and parties arrive at trade policy positions in anticipation of an election campaign. Considering that the sponsorship of policies that benefit a narrow client group may not reflect highly on a politician (although it may add to his campaign funding), there is an incentive for politicians to enact trade policies which benefit the client group, but which are less than transparent to the voting public.

The deGorter and Tsur chapter is a good example of the new political economy approach applied to agriculture. In their model, two factors are important for determining the recipient and amount of government support: relative income between groups, and the redistribution of support and income within groups. It is hypothesized that governments increase political support given to them by exploiting differences in the marginal utility of income of interest groups. Voter behavior within a group (or the intensity of support for the government) depends on the redistribution of income within the group. As Honma notes, although this model may help explain agricultural support in countries like Britain and West Germany, its applicability to other countries, especially less developed countries (LDC's), may be limited. This type of observation implies that structural political economy models should best be interpreted in terms of particular problems for which they have been developed. Their generality may indeed be limited.

Moore suggests a number of research areas which merit further attention. Legislative mechanisms need to be analyzed in addition to electoral processes. Here the emphasis is on multi-issue rather single-issue decision making. However, as Roningen has noted in this volume, legislative markets are typically "thin" markets, and economists have not had much success in predicting outcomes in such markets. Much effort will be needed in examining vote-trading and log rolling in specific contexts. Predicted outcomes will likely differ from the "policy equilibria" which characterize most of the literature up to this point. The analysis of "policy games" will, for the most part, be dynamic; that is, in the nature of a repeated game.

Bureaucratic decision making needs to be analyzed as well. The emphasis is on how laws determined in political and legislative markets are implemented, and on how administered protection procedures (such as countervailing and antidumping duties) are actually carried out. Messerlin (1983) has already hypothesized that the jurisdiction of a bureaucracy determines its institution attitude toward trade policy. The bureaucracy supports the factors of production or industry that it supervises. The support for protection will likely be in the form of nontariff barriers that are complex, non-transparent, and in need of bureaucratic administration. (However, it is likely as well that agencies with broad responsibilities will be more supportive of liberalized trade.) Associated with the analysis of bureaucracies is the question of how firms decide which type of protection track (political or bureaucratic) they will pursue.

Political economy analysis needs to consider relationships between governments of rival trading partners. There always exists the possibility of foreign retaliation to trade policy measures. Lobbyists for foreign countries now play a very active role in the domestic legislative process. Also, the existence of "Super 301" which is meant to be used to open foreign markets through threats of U.S. retaliation to "unfair" trading practices will provide interesting interactions with foreign governments and firms alike.

Empirical Research

Ray presents a survey of results from recent political economy empirical research. There is a dual focus to this research. The first focus area emphasizes special interest groups. Here there are two sets of issues. First, what are the determinants of the structure of protection across industries? Second, what determines the level of protection to a particular industry? The second focus is the political process. What are the characteristics of voters and political representatives that act as determinants in influencing and/or directing administrative decisions on trade cases?

There are a number of research results which have emerged. Tariff protection is typically associated with industries in which the country has a long-standing comparative disadvantage in trade. Also, nontariff trade barriers (NTB's) are often used to supplement tariff protection in these industries as well. In the United States, these industries are low-skill and labor intensive industries like consumer goods, textiles, and processed agricultural products. (As an aside, Ray notes that tariff cuts resulting from the General Agreement on Trade and Tariffs (GATT) rounds have been biased away from commodity groups in which LDC's have the greatest potential for exports to developed countries.) NTB's are widely used to provide trade restrictions in traditionally less protected industries. These are industries which typically produce standardized products using capital intensive methods of production (like automobiles). There has been a proliferation in the use of NTB's recently, and they tend to be centered in less concentrated industries.

Voting on protectionist legislation usually reflects the economic interests of legislators' constituents. This pattern is especially true for the House of Representatives, and less so for the Senate. Important variables identified in predicting protectionist voting are special interest campaigns, union strength within the legislator's state, and the state unemployment rate. Export interests, as well as ideological bents, are relatively unimportant as determinants of voting patterns.

Administrative protection measures such as countervailing duties and antidumping provisions are increasingly important. Firms, usually in less concentrated industries, make use of the administrative track because of reduced organizational costs for success. An implication is that calls for tariffication of NTB's seem particularly naive, given that firms receiving NTB support through administrative procedures know of their inability to sustain support for themselves through tariffs.

All in all, the empirical research on the costs and benefits of protection (especially that for the automobile, lumber, and steel industries) shows that industry interests usually triumph over the interests of consumers. Consumer costs are usually many times the value of job losses prevented in a given industry. This result should help to temper the implications of the new trade strategic trade theory which emphasizes potential gains to the general economy from strategic intervention of government in trade.

There are a number of questions needing to be answered. For instance, how is protection in one country matched by trading partners? What are third party responses to bilateral trade agreements? What are the domestic resource costs of protectionist policies? What is the distribution of firms and industries among forms of trade restraints? What has been the evolution of innovations in protection, including the who and the why? What are the linkages between macroeconomic policies and industry specific protection? The effect of the macroeconomy, especially the implications of flexible exchange rates and business cycle fluctuations, seems particularly relevant for agriculture (Roningen).

Additional Perspective: Multinational Corporations

One facet of new oligopoly trade theory not dealt with in this volume and deserving of more attention is the phenomena of multinational corporations. Trade in capital is increasingly displacing trade in goods. Ohmae (1987) has argued that the measure of a nation's penetration in global markets should be the sum of a nation's total of goods exported and of goods produced abroad by subsidiaries of companies from that nation. Consider that in 1984 the United States exported $25.6 billion to Japan, and the value of made-in-Japan American goods was $43.9 billion, for a total sum of $69.5 billion. At the same time, the United States imported $56.8 billion from Japan, and purchased $12.8 billion of made-in-U.S. Japanese goods, for a total of $69.6 billion. These totals are remarkably close (at least for 1984). When considered on a per capita basis, the Japanese spent twice as much on American products as Americans spend on Japanese products. Lipsey and Kravis (1986) report that although the export share of production by U.S. based parent companies has declined since 1966, gains by majority owned affiliates have more than offset the drop. The point is that the world is becoming an increasingly unified marketplace where trade in capital may be replacing trade in goods.

New theories paralleling those of the new trade theory have been developed to explain multinational activity. Helpman (1984) has emphasized that multinational enterprise activity is generated by multi-plant scale economies, or by ownership of factors of production that generate the scale economies. Locational considerations depend on relative factor endowments and technological parameters. Ethier (1986) has linked multinational activity to the economics of information. He maintains that if there exists the necessity of exchanging a large volume of diverse information between "upstream" and "downstream" activities across national borders, then the likelihood of vertical integration of these activities within the multinational framework is high.

The insights from the multinational theory seem particularly relevant for U.S. food processors. Research by Seigle and Handy reported in Farmline (Martinez 1989) shows that U.S. food processors tend to produce finished consumer food products in foreign plants rather than export from the United States. Out of 62 food processing firms surveyed, exports constituted only 2.9 percent of their sales. The value of exports of these firms totaled only $3.4 billion for 1986-87. Sales from foreign operations accounted for 19 percent of total processed food sales for the 62 firms. Although foreign trade policies have some role in explaining this phenomenon, the choice of production site is influenced much more by transportation costs and especially nation-specific taste preferences.

Ethier's theory seems particularly relevant in this context. Due to differences in food taste preferences, it is difficult to service foreign markets through the export of products developed for the U.S. market. It makes more sense to locate plants close to the market in order to develop and maintain the product to suit localized tastes. The neo-Hoteling view of horizontal differentiation (that is, the ideal variety approach described by MacLaren) may be the most appropriate, although the implications are for reduced trade rather than increased trade.[1]

Multinational firm theory is applicable to other agricultural areas as well. The largest grain trading firms purchase, as well as sell, grain from abroad. Agricultural machinery firms have increasingly shifted production of certain lines of production abroad to be closer to the largest markets for those products. They then import the lines produced abroad to the United States to meet the smaller U.S. demand. U.S. agricultural chemical firms operate abroad as well. A combination of multinational and new trade theory may be useful in exploring these industries and markets.

Conclusion

To sum up, we ask again, "What do the new trade and political economy approaches have to offer the agricultural trade economist?" Another way to ask this question is to think of a scale of 0 to 10 where 0 represents complete ignorance about agricultural trade and 10 represents truth and understanding. Traditional theory helped us move up that scale to our current position (at 6?). What is the potential for the new approaches to move us a little closer to 10? Our assessment of the authors' comments is that these newer concepts do in fact provide new insights, but they come at a cost. Substantially more data will be needed and more complex models will be needed to quantify relationships and to test hypotheses that are derived from the new theory. Thus the practical trade economist has to judge whether the expected benefits (i.e., the rewards from moving toward 10 on our scale) exceed the expected costs. Our hope is that risk-seeking agricultural trade researchers will charge ahead and further explore these concepts.

Notes

1. There may be products, however, that fit the prediction of increased trade. Seigle and Handy in Farmline (Martinez 1989) mention almonds; canned fruits, juices, and vegetables; and jellies and jams. Processed tobacco products (cigarettes) probably fit in this category as well.

References

Ethier, W.J. 1986. The Multinational Firm. *The Quarterly Journal of Economics.* Vol CI:805-33.

Helpman, E. 1984. A Simple Theory of International Trade with Multinational Corporations. *Journal of Political Economy.* 92:451-71.

Honma, M. and Y. Hayami. 1986. Structure of Agricultural Protection In Industrialized Countries. *Journal of International Economics.* 20:115-29.

Krugman, P.R. 1981. Intraindustry Specialization and the Gains from Trade. *Journal of Political Economy.* 89:959-73.

Krugman, P.R. 1987. Is Free Trade Passe? *The Journal of Economic Perspectives.* 1:131-44.

Lipsey, R.E. and I.B. Kravis. 1986. The Competitiveness and Comparative Advantage of U.S. Multinationals, 1957-1983. National Bureau of Economic Research.

Martinez. 1989. U.S. Food Processors Expanding Overseas. *Farmline.* p.4-7.

Messerlin, P. 1983. *Bureaucracies and the Political Economy of Protection: Reflections of a Continental European.* World Bank Staff Working Papers, no. 568.

Ohmae, K. 1987. *Beyond National Border*. Homewood, IL: Dow Jones-Irwin.
Olson, M. 1965. *The Logic of Collective Action*. Cambridge, Ma: Harvard Univ.
  Press.

Contributors

Giovanni Anania
Department of Economics
University of Calabria
87036 Arcavacata di Rende (CS)
Italy

Nicole Ballenger
Economic Research Service
U.S. Department of Agriculture
1301 NY Ave., N.W.
Washington, DC 20005 U.S.A.

Colin A. Carter
Department of Agricultural
    Economics
University of California
Davis, CA 95616 U.S.A.

Harry de Gorter
Department of Agricultural
    Economics
Cornell University
Ithaca, NY 14853 U.S.A.

Stephen Haley
Agriculture and Trade Analysis
    Division
Economic Research Service
U.S. Department of Agriculture
1301 New York Ave., N.W.
Washington, DC 20005 U.S.A.

Masayoshi Honma
Department of Economics
3 5 21 Midori Otaru
Otaru University of Commerce
Hokkaido, Japan 047

Kala Krishna
Department of Economics
Harvard University
Cambridge, MA 02138 U.S.A.

Donald MacLaren
School of Agriculture and Forestry
The University of Melbourne
Parkville, Victoria 3052
Australia

Alex F. McCalla
Department of Agricultural
   Economics
University of California
Davis, CA 95616 U.S.A.

Michael O. Moore
Department of Economics
The George Washington
   University
2201 G. Street, N.W.
Washington, D.C. 20052 U.S.A.

Philip L. Paarlberg
Department of Agricultural
   Economics
Krannert School of Management
Purdue University
W. Lafayette, IN 47907 U.S.A.

Edward John Ray
Department of Economics
The Ohio State University
410 Arps Hall
1945 North High Street
Columbus, OH 43210-1172 U.S.A.

J. David Richardson
Economics Department
University of Wisconsin
Madison, WI 53706 U.S.A.

Vernon Roningen
Economic Research Service
U.S. Department of Agriculture
1301 New York Ave., N.W.
Washington, DC 20005 U.S.A.

Jerry Sharples
Agriculture and Trade Analysis
   Division
Economic Research Service
U.S. Department of Agriculture
1301 New York Ave., N.W.
Washington, DC 20005 U.S.A.

Jerry G. Thursby
Krannert School of Management
Purdue University
W. Lafayette, IN 47907 U.S.A.

Marie C. Thursby
Krannert School of Management
Purdue University
W. Lafayette, IN 47907 U.S.A.

Yacov Tsur
Department of Agricultural and
   Applied Economics
University of Minnesota
St. Paul MN 55108

Michele M. Veeman
Department of Rural Economy
University of Alberta
Edmonton, Alberta
Canada T6G 2H1

Harald von Witzke
Center for International Food and
   Agricultural Policy
Department of Agricultural and
   Applied Economics
University of Minnesota
1994 Buford Avenue
St. Paul, MN 55108 U.S.A.

Index

Printed and bound by CPI Group (UK) Ltd, Croydon, CR0 4YY

23/10/2024

01778240-0002